9

The Navstar Global Positioning System

Tom Logsdon

VNR VAN NOSTRAND REINHOLD
——————————— New York

Copyright © 1992 by Van Nostrand Reinhold

Library of Congress Catalog Card Number 92-10521
ISBN 0-442-01040-0

Printed in the United States of America

Van Nostrand Reinhold
115 Fifth Avenue
New York, New York 10003

Chapman and Hall
2-6 Boundary Row
London, SE 1 8HN, England

Thomas Nelson Australia
102 Dodds Street
South Melbourne 3205
Victoria, Australia

Nelson Canada
1120 Birchmount Road
Scarborough, Ontario M1K 5G4, Canada

16 15 14 13 12 11 10 9 8 7 6 5 4 3 2 1

Library of Congress Cataloging-in-Publication Data

Logsdon, Tom, 1937–
    The Navstar global positioning system / by Tom Logsdon.
       p.   cm.
    Includes bibliographical references and index.
    ISBN 0-442-01040-0
    1. Artificial satellites in navigation—United States.   2. Global
Positioning System.   I. Title.
TL798.N3L64   1992
629.04′5—dc20   92-10521
                                                                  CIP

Dedication

To Chad, who someday may be using the modulated signals from the Navstar satellites to guide his noisy trailer truck along the Interstate.

Contents

Preface

During the Persian Gulf War a group of American soldiers scooped up a new recruit at Rijaid Airport, then drove him, with blackened headlights, directly across miles of tractless desert sand. Squinting toward the horizon, he could see almost nothing when suddenly the driver mashed on the brakes, gave him a quick salute, and instructed him to step out into the darkness. As his boots sank into the sand, he was stunned to realize that he was only a few feet away from the flap of his tent. Before setting out, the driver had keyed the tent's coordinates into a Navstar receiver, so it could guide him back again.

No one knows exactly how many Navstar receivers ended up serving coalition forces along the Persian Gulf because mothers and fathers—and sweethearts, too—located a few stray units on the shelves of marine supply houses, plunked down their money, and express mailed them to their loved ones in the Persian Gulf.

A few resourceful soldiers called stateside suppliers long distance, then used their credit cards to order receivers, many of which arrived in Saudi Arabia a day or two later aboard commercial jetliners. By the time the ground war finally started, 4,000 to 7,000 Navstar receivers were clutched in the hands of grateful American soldiers. They were used to guide fuel-starved airplanes for linkups with aerial tankers, to pull in air strikes against enemy emplacements, to guide mess trucks toward hungry troops, and to vector Special Forces units in their muffled dune buggies deep behind enemy lines.

A few enterprising military engineers learned how to follow meandering goat trails so they could locate underground springs where the goats watered themselves. They then used their hand-held Navstar receivers to record the precise coordinates of each spring, thus insuring fresh water supplies for onrushing troops.

Unlike most of its predecessors, which rely on ground-based transmitters to fix the user's position, the Navstar GPS employs orbiting satellites. From their high-altitude vantage points 11,000 nautical miles above the earth, the Navstar satellites broadcast precise, reliable, and continuous navigation signals to a worldwide class of users. Military dollars pay for the satellites, but their signals are available free of charge to anyone, anywhere, who decides to use them.

Navstar receivers, many as small and compact as pocket calculators, are available from 50 different manufacturers. Most of them are simple and easy to operate, and, even under worst-case conditions, their average accuracy is 50 to 100 feet. The least expensive civilian models cost less than $1,500 each.

This book is targeted toward intelligent Navstar users who are not already world-class experts on the many facets of Navstar navigation. Its discussions are intended to help novices and professionals alike learn all they need to know to evaluate the potential of this exciting new spaceborne system, select fruitful applications, choose an appropriate Navstar receiver, and obtain maximum benefit from its use.

A book is invariably a team effort, and this one was no exception. Many cooperative individuals contributed toward its successful completion. I would like to thank, in particular, the 3,000 practicing professionals who have, over the past few years, attended my many broad-ranging short courses on Navstar navigation. Their many helpful comments and suggestions have strongly shaped its contents, and, often, their penetrating questions have sent me back to the library for fresh research.

Those short courses have, incidentally, been offered in more than a dozen American cities and in 16 different countries on four continents. Despite her untiring efforts, my wife Cyndy was not able to accompany me to each and every one of those foreign countries. She has, however, made many invaluable contributions in structuring, shaping, and pruning the final manuscript.

As usual, my hard-working typist, Elda Stramel, kept big stacks of flawless pages pouring from her trusty typewriter. For years, Elda has been a treasured asset, who has helped me put together more than a dozen books. My literary agent, Jane Jordan Browne, displayed extraordinary tenacity and patience in helping get this project off the ground. Both she and her husband have earned my long-lasting gratitude. So have Lauren Weinnerod and Steve Chapman, my editors at Van Nostrand Rinehold, who worked so diligently to shepherd the manuscript through its various stages of production.

Most of the art work was handled by Lloyd Wing and Anthony Vega who, long ago, managed to master the intricacies of the MacIntosh computer. Their many long nights, nose to the grindstone, are greatly appreciated. So are the efforts of my consultant, Dr. Jim Haffner, who has long provided me with ample technical support. Dr. Haffner knows more about *everything* than most people know about *anything*. But, despite his remarkable expertise, he somehow manages to display total cordiality toward mere mortal engineers.

Numerous individuals caught and corrected minor errors in the manuscript during its various stages of production. Any errors that managed to elude them, however, are the sole responsibility of the author. This responsibility is being accepted with only vanishingly small enthusiasm. Unfortunately, few qualified candidates have stepped forward to help shoulder the blame.

Tom Logsdon
Seal Beach, California
1992

1

The Science of Navigation

Mankind's earliest navigational experiences are lost in the shadows of the past. But history does record a number of instances in which ancient mariners observed the locations of the sun, the moon, and the stars to help direct their vessels across vast, uncharted seas. Bronze Age Minoan seamen, for instance, followed torturous trade routes to Egypt and Crete, and, even before the birth of Christ, the Phoenicians brought many shiploads of tin from Cornwall. Twelve hundred years later, the Vikings were probably making infrequent journeys across the Atlantic to settlements in Greenland and North America.

How did these courageous navigators find their way across such enormous distances in an era when integrating accelerometers and hand-held receivers were not yet available in the commercial marketplace? Herodotus tells us that the Phoenicians used the Pole Star to guide their ships along dangerous journeys, and Homer explains how the wise goddess instructed Odysseus to "keep the Great Bear on his left hand" during his return from Calypso's Island. Another account in the Acts of the Apostles indicates that, in biblical times, navigators used the stars and the sun to distinguish between north, south, east, and west.

Eventually, the magnetic compass reduced mankind's reliance on celestial navigation. One of the earliest references to compass navigation was made in 1188, when Englishman Alexander Neckam published a colorful description of an early version consisting of "a needle placed upon a dart which sailors use to steer when the Bear is hidden by clouds." Eighty years later the Domican friar, Vincent of Beauvais, explained how daring seamen, whose boats were deeply shrouded in fog, would "magnetize the needle with a loadstone and place it through a straw floating in water." He then went on to note that "when the needle comes to rest, it is pointing at the Pole Star." The sextant, which was developed and refined over several centuries, made Polaris and its celestial neighbors considerably more useful to navigators on the high seas. When the sky was clear, this simple device—which employs adjustable mirrors to measure the elevation angles of stellar objects with great precision—could be used to nail down the latitude of a ship so that ancient navigators could maintain an accurate east-west heading. However, early sextants were largely useless for determining longitude because reliable methods for measuring time aboard ship were not yet available.

The latitude of a ship equals the elevation of the Pole Star above the local horizon, but its longitude depends on angular measurements and the precise time. The earth spins on its axis 15 degrees every hour; consequently, a one-second timing error translates into a longitudinal error of 0.004 degrees—about 0.25 nautical miles at the equator. The best seventeenth-century clocks were capable of keeping time to an accuracy of one or two seconds over an interval of several days, when they were sitting on dry land. But, when they were placed aboard ship and subjected to wave pounding, salt spray, and unpredictable variations in temperature, pressure, and humidity, they either stopped running entirely or else were too unstable to permit accurate navigation.

To the maritime nations of seventeenth century Europe, the determination of longitude was no mere theoretical curiosity. Sailing ships by the dozens were sent to the bottom by serious navigational errors. As a result of these devastating disasters caused by inaccurate navigation, a special act of Parliament established the British Board of Longitude, a study group composed of the finest scientists living in the British Isles. They were ordered to devise a practical scheme for determining both the latitude and the longitude of English ships sailing on long journeys. After a heated debate, The Board offered a prize of £20,000 to anyone who could devise a method for fixing a ship's longitude within 30 nautical miles after a transoceanic voyage lasting six weeks.

One proposal advanced by contemporary astronomers would have required that navigators take precise sightings of the moons of Jupiter as they were eclipsed by the planet. If practical trials had demonstrated the workability of this novel approach, ephemeris tables would have been furnished to the captain of every flagship or perhaps every ship of the British fleet. The basic theory was entirely sound, but, unfortunately, no one was able to devise a practical means for making the necessary observations under the rugged conditions existing at sea.

However, in 1761, after 47 years of painstaking labor, a barely educated British cabinetmaker named John Harrison successfully claimed the £20,000 prize, which in today's purchasing power would amount to about $1 million. Harrison's solution centered around his development of a new shipboard timepiece, the marine chronometer, which was amazingly accurate for its day. On a rocking, rolling ship in nearly any kind of weather, it gained or lost, on average, only about one second per day. Thus, under just about the worst conditions imaginable, Harrison's device was nearly twice as accurate as the finest land-based clocks developed up to that time.

For the next two centuries, precise timing measurements from marine chronometers coupled with sextant sightings of planets and stars represented the only reliable means of determining a ship's position in unfamiliar waters. The sextant is still widely used today on the ground, at sea, and even in space, but modern radionavigation techniques provide much more practical and efficient methods for finding both longitude and latitude with the desired precision.

During World War II, ground-based radionavigation systems came into widespread use when military commanders in the European theater needed to vector their bombers toward specific targets deep in enemy territory. Both Allied and Axis researchers soon learned that ground-based transmitters could provide reasonably accurate navigation within a limited coverage regime.

In the intervening years America and various other countries have operated a number of ground-based radionavigation systems. Many of them—Decca, Omega, Loran—have been extremely successful. But in recent years, American and Russian scientists have been moving their navigation transmitters upward from the surface of the earth into outer space. There must be some compelling reason for installing navigation transmitters aboard orbiting satellites. After all, it costs something like $40 million dollars to construct a navigation satellite and another $40 million to launch it into space. Moreover, at least a half-dozen orbiting satellites are needed for a practical spaceborne radionavigation system. Later in this chapter you will learn why space is such an attractive location for navigation transmitters. But first let's pause to define a few fundamental concepts and briefly describe some of the more common navigation techniques now being used.

What Is Navigation?

Navigation can be defined as *the means by which a craft is given guidance to travel from one known location to another.* Thus, when we navigate, we not only determine *where we are*, we also determine *how to go from where we are to where we want to be.*

Five practical methods of navigation are in widespread use:

1. Piloting
2. Dead reckoning
3. Celestial navigation
4. Inertial navigation
5. Electronic or radionavigation

Piloting, which consists of fixing a craft's position with respect to familiar landmarks, is the simplest and most ancient method of navigation. In the 1920s bush pilots often employed piloting to navigate from one small town to another. Such a pilot would fly along the railroad tracks out across the prairie, swooping over isolated farmhouses along the way. Upon arrival at village or town, he would search for a water tower with the town's name painted in bold letters to make sure he had not overshot his intended destination.

Dead reckoning is a method of determining position by extrapolating a series of measured velocity increments. In 1927 Charles Lindbergh used dead reckoning when he flew his beloved *Spirit of St. Louis* on a 33-hour journey from Long Island to Le Bourget Field outside Paris. Incidentally, Lindbergh hated the name. Originally it had been called *"ded reckoning"* for *"deduced reckoning,"* but newspaper reporters of the day could never resist calling it "dead reckoning" to remind their readers of the many pilots who had lost their lives attempting to find their way across the North Atlantic.

Celestial navigation is a method of computing position from precisely timed sightings of the celestial bodies, including the stars and the moon. Primitive celestial navigation techniques date back thousands of years, but celestial navigation flourished anew when cabinetmaker John Harrison constructed surprisingly accurate clocks for use in conjunction with sextant sightings aboard British ships sailing on the high seas. The uncertainty in a celestial navigation measurement builds up at a rate of a quarter of a nautical mile for every second timing error. This cumulative error arises from the fact that the earth rotates to displace the stars along the celestial sphere.

Inertial navigation is a method of determining a craft's position by using integrating accelerometers mounted on gyroscopically stabilized platforms. Years ago navigators aboard the Polaris submarine employed inertial navigation systems when they successfully sailed under the polar icecaps.

Electronic or *radionavigation* is a method of determining a craft's position by measuring the travel time of an electromagnetic wave as it moves from transmitter to receiver. The position uncertainty in a radionavigation system amounts to at least one foot for every billionth of a second timing error. This error arises from the fact that an electromagnetic wave travels at a rate of 186,000 miles per second or one foot in one-billionth of a second.

According to The Federal Radionavigation Plan published by the United States government, approximately 100 different types of domestic radionavigation systems are currently being used. All of them broadcast electromagnetic waves, but the techniques they employ to fix the user's position are many and varied. Yet, despite its apparent complexity, radionavigation can be broken into two major classifications:

1. Active radionavigation
2. Passive radionavigation

A typical *active radionavigation* system is sketched in Figure 1.1. Notice that the navigation receiver fixes its position by transmitting a series of precisely timed pulses to a distant transmitter, which immediately rebroadcasts them on a different frequency. The slant range from the craft to the distant transmitter is established by multiplying half the two-way signal travel time by the speed of light.

In a *passive radionavigation* system (see Figure 1.1), a *distant* transmitter sends out a series of precisely timed pulses. The navigation receiver picks up the pulses, measures their signal travel time, and then multiplies by the speed of light to get the slant range to that transmitter.

A third navigational approach is called *bent-pipe navigation*. In a bent-pipe navigation system a transmitter attached to a buoy or a drifting balloon broadcasts a series of timed pulses up to an orbiting satellite. When the satellite picks up each timed pulse, it immediately rebroadcasts it on a different frequency. A distant processing station picks up the timed pulses and then uses computer processing techniques to determine the approximate location of the buoy or balloon.

ACTIVE AND PASSIVE RADIONAVIGATION SYSTEMS

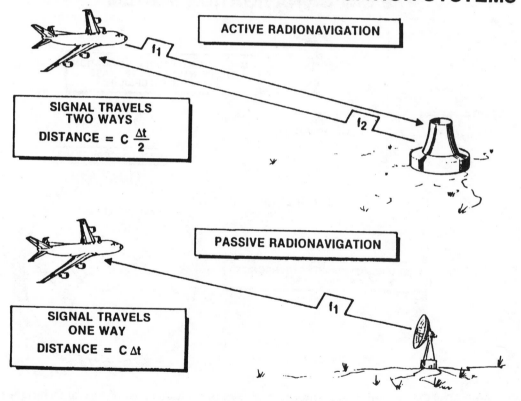

Figure 1.1 Most radionavigation systems determine the user's position by measuring the signal travel time of an electromagnetic wave as it travels from one location to another. In active radionavigation the timed signal originates on the craft doing navigating. In passive radionavigation it originates on a distant transmitter.

A Typical Ground-Based Radionavigation System

Omega provides us with an instructive example of how a ground-based radionavigation system operates. The eight Omega transmitters, which are dispersed around the globe, are phase-locked-looped together. This means that the electromagnetic carrier waves originating from the various transmitters are always in phase. Phase coherence is accomplished by rigging each transmitter to monitor the carrier waves from its neighbors and then making subtle adjustments to its own transmissions. Precise atomic clocks at each site help to maintain the accuracy and integrity of the phase lock loop.

A shipborne receiver picking up the carrier waves from two of the Omega transmitters will observe a phase-difference-of-arrival because the two carrier waves travel along two different path lengths to reach the receiver. If the two separate carrier waves could be displayed simultaneously on an oscilloscope, the phase difference between them would become readily apparent (see Figure 1.2). Each discrete phase displacement is associated with a

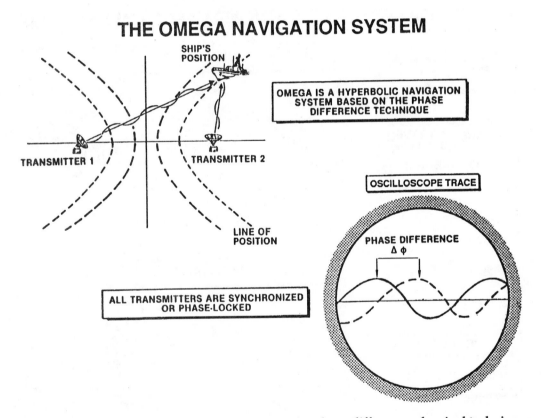

THE OMEGA NAVIGATION SYSTEM

Figure 1.2 The Omega radionavigation system uses phase-difference-of-arrival techniques to determine the user's position. When a receiver picks up the carrier waves from two of the eight widely dispersed Omega transmitters, the measured phase difference fixes the receiver on a specific hyperbolic line of position. By picking up similar signals from two other Omega transmitters, the receiver can fix its longitude and latitude at the intersection of two hyperbolas.

specific path-length difference that, in turn, fixes the user on a particular hyperbolic line of position. By picking up similar electromagnetic waves from two other transmitters, the navigation receiver fixes the ship on a second hyperbolic line of position. Once this has been accomplished, the user is known to lie at the intersection of the two hyperbolas.

When the carrier waves from two different Omega transmitters are displayed on an oscilloscope, it is not obvious how much they differ. They could, for instance, differ by a half wavelength, one and a half wavelengths, two and a half wavelengths, and so on. Any radionavigation system that suffers from this shortcoming is said to exhibit "lane ambiguities."

The very low frequency carrier waves for the Omega system are approximately 16 miles long. So it may seem obvious that any user of the Omega system must be able to estimate his or her initial location to within ± 8 miles in order to benefit from Omega navigation. However, the designers of the Omega realized that lane ambiguities could be a problem, so they designed their transmitters to operate on four different frequencies instead of only one. Each frequency produces its own particular lane ambiguity, and, when they are all combined, the overall lane ambiguity turns out to equal 72 miles.

The eight Omega transmitters are located at Norway, Liberia, Hawaii, North Dakota, Diego Garcia, Argentina, Australia, and Japan. They operate sequentially, with each one transmitting on each of four navigation frequencies for approximately one second. Each transmission is followed by a 0.2-second "guard band," during which the transmitter is silent. The eight different transmission intervals are staggered in time with respect to one another, so there is no overlap between the four different navigation frequencies eminating from the various Omega transmitters. At each transmitter, the four navigation frequencies, taken together, are active approximately one-half the time. During the rest of the time (guard bands excluded), each transmitter is broadcasting its "identification frequency," which uniquely identifies it to the various users of the system.

The Advantages of Space-based Transmitters

Ground-based radionavigation transmitters have been providing reliable navigation coverage to a worldwide class of users for more than 50 years. So why are today's ground-based transmitters being replaced by navigation transmitters positioned in space? The engineers who design a ground-based radionavigation system have essentially two choices when they are selecting its transmission frequency, neither of which provides entirely satisfactory results. If they select certain specific very low frequency transmission waves, they can achieve "wave-form ducting," in which the carrier waves reflect off the ionosphere. This broadens the coverage area, so a small number of transmitters can provide coverage for a substantial fraction of the globe.

The Omega ground-based system, for instance, provides essentially global coverage with only eight widely scattered transmitters. Unfortunately, Omega and other similar systems, provide rather inaccurate navigation because their carrier waves cannot be modulated with much useful information, and because their signals reflect off the ionosphere, which experiences large variations in height and composition. The Omega ground-based system, for instance, uses carrier waves that are about 16 miles long. Consequently, the navigation solutions it provides are rather inaccurate. Its error typically amounts to about 4 nautical miles in longitude and latitude at the 2σ probability level (95 percent of the time).

On the other hand, if the designers of a radionavigation system choose to employ high-frequency carrier waves, the resulting navigation solutions will likely be far more accurate. Unfortunately, high-frequency carrier waves punch through the earth's ionosphere, so they provide only line-of-sight coverage in a small, local area.

Even with a transmission tower 300 feet high, the circular line-of-sight coverage area is typically only about 40 miles in diameter. Worldwide coverage would thus require thousands of ground-based (and sea-based) transmitters scattered around the globe. Fortunately, there is a solution to this dilemma. Employ high-frequency carrier waves broadcast from transmitters high above the earth in outer space. From its high-altitude vantage point, a space borne transmitter can cover a substantial fraction of the globe with high-frequency carrier waves that penetrate the earth's ionosphere from the top down, to provide users with global coverage and highly accurate navigation.

Specifically, the Navstar satellites whiz through space at an altitude of 10,898 nautical miles, where each satellite gains direct line-of-sight access to 42 percent of the globe. These cleverly designed satellites broadcast high-frequency carrier waves at 1227.6 and 1575.42 megahertz that are 9.6 and 7.5 inches long, respectively. Their unobstructed view of the ground below and their high frequency transmissions greatly enhance the accuracy and coverage of the Navstar constellation.

The Transit Navigation Satellites

America's first family of radionavigation satellites, The Transit Navigation System, worked as advertised, but it did not reap all the benefits from the high-altitude vantage point in space. The Transit Navigation satellites are launched into polar "bird cage" orbits 580 nautical miles above the earth, a low-altitude orbit that carries them directly over the north and south poles. Five or six Transit Navigation satellites are usually orbiting the earth at any given moment. They provide global, but intermittent, coverage for thousands of users on the ground below.

As the Transit satellites travel around their orbits, they transmit two continuous tones that to ground-based user sets appear to experience systematic "Doppler shift" frequency changes as the satellite swings across the sky. By observing the Doppler shift over a substantial fraction of the satellite's orbit, the user set can determine its current position with a fair degree of accuracy.

Doppler shift frequency changes are created when a stationary observer picks up a continuous wave (tone) from a moving transmitter. The plaintive whistle emitted by a locomotive can be used to illustrate how the Doppler shift from the Transit Navigation satellites can be used to determine the user's position. Suppose you are standing next to a railroad track when a high-speed train passes by. As the train approaches, the waves from its whistle will be compressed to produce a higher pitch. When the train recedes, the waves will be stretched out to produce a lower pitch than would otherwise be observed. If you construct a graph of frequency versus time, it will turn out to be a step function if you are standing right beside the track.

But, if you move back away from the track and construct a similar graph of Doppler shift versus time, it will turn out to be a gentle S-shaped curve. Its gentle curvature arises from the fact that the systematic shift in pitch is created by the component of velocity along your instantaneous line-of-sight vector to the train. The exact shape of your Doppler shift curve can provide an estimate of how far you are away from the track. Moreover, if the train has a published schedule and it broadcasts the exact time (modulated on its whistle), this timing information can be used to pinpoint your lateral location along the track.

The Transit Navigation System employs conceptually similar position-fixing techniques. Of course, it works with electromagnetic waves rather than sound. A Transit (SatNav) receiver measures the shape of the Doppler shift curve as a satellite travels along its orbit from horizon to horizon. Then, in essence, it executes a curve-fitting routine to determine the shape of the Doppler shift curve, which indicates how far it is from the satellite's ground trace. A data stream broadcast by the Transit Navigation satellite allows the receiver to determine the satellite's orbit and the exact time. This information, together with the precise contours of the Doppler shift curve, allows users on the ground or at sea to obtain one fairly accurate position estimate each time a Transit Navigation satellite passes from horizon to horizon. This typically takes 10 or 15 minutes.

Orbital updates and clock correction factors for the Transit satellites are provided by the Navy's navigation experts at Point Mugu, California. The orbital ephemeris constants and the clock correction factors they send up to the satellites are broadcast back down to the receivers on the ground to aid their navigation solutions.

Gravity Gradient Stabilization

To provide global navigation coverage, the Transit satellites must constantly blanket the full disk of the earth with radio frequency transmissions. Consequently, the navigation antennas onboard the satellites must be continuously oriented toward the center of the earth. This is accomplished by a clever altitude control system that uses no propellants, called *gravity gradient stabilization.*

A Transit Navigation satellite employs gravity gradient stabilization by flying into orbit, swiveling into an earth-seeking orientation, and then deploying a 50-foot telescoping boom vertically upward away from the earth (see Figure 1.3). The pull of gravity is always stronger on the lower edge of the boom than on its upper edge. So, if the satellite begins to drift away from its desired vertical orientation, the difference in gravity (the gravity-gradient) will create a restoring torque to nudge it back into a vertical orientation. Actually, a disturbance causes the satellite to oscillate about the vertical, but electrical hysteresis loops can be used to damp out any oscillations to achieve a continuous earth-seeking orientation. No rocket propellants are required because the hysteresis loops are powered by electricity from the satellite's solar cells.

THE TRANSIT SATELLITES ARE LAUNCHED INTO POLAR "BIRDCAGE" ORBITS

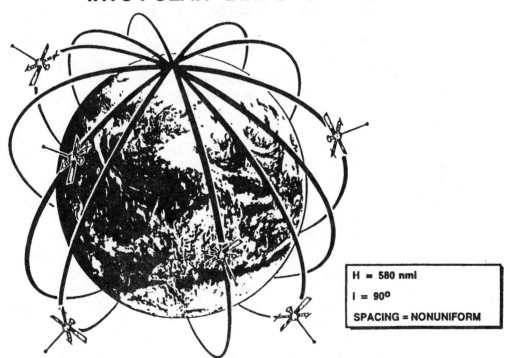

| H = 580 nmi |
| I = 90° |
| SPACING = NONUNIFORM |

Figure 1.3. Each Transit Navigation satellite is launched into a polar "bird cage" orbit 580 nautical miles above the earth. When the satellite arrives in space, a long telescope boom is extended vertically upward away from the earth to take advantage of the gravity-gradient, the variation in gravity along the telescoping boom. The gravity-gradient helps the spacecraft maintain a precise earth-seeking orientation without using propellants.

Disturbance Compensation Systems

To achieve improved spacecraft autonomy (independence from frequent ground updates), the advanced Nova versions of the Transit Navigation satellites employ a so-called *disturbance compensation system*—a clever design feature that helps eliminate the effects of the orbital perturbations caused by solar radiation pressure and drag with the earth's upper atmosphere.

A disturbance compensation system employs a compact proof mass floating inside a hollow cavity near the center of gravity of the spacecraft. The proof mass is inside the spacecraft, so it is shielded from solar radiation pressure and atmosphereic drag. Thus, it travels around the earth along a free-fall trajectory as though these perturbations did not exist. Of course, the spacecraft itself *is* affected, so, if nothing is done, the proof mass will eventually slam into the walls of the hollow cavity. This is prevented by setting up a feedback control loop to monitor the location of the proof mass, and firing thrusters mounted on the outside of the spacecraft to recenter the proof mass whenever it begins to drift. With these adjustments the spacecraft flies in formation with the proof mass inside; therefore, it travels along the same path it would follow if drag and solar radiation pressure did not exist.

Actually, the advanced Nova satellites employ a simpler version of the disturbance compensation system just described. They are rigged with a polished metal cylinder that slides back and forth on a polished metal rod. This eliminates practically all of the atmospheric drag (which is pushing the satellite in the direction opposite to its orbital motion). Any solar radiation pressure forces that happen to lie along the instantaneous velocity vector are also eliminated.

The simpler, more predictable orbits that result provide increased autonomy for the Transit Navigation satellites, so their orbital elements need not be updated nearly as often. This saves manpower for the ground crews tending the satellites.

Compensating for Ionospheric Delays

The navigation signals streaming down to the ground from the Transit Navigation satellites are bent and slowed down as they pass through the earth's ionosphere. The resulting time delay depends on the number of ions and free electrons that lie along the line-of-sight vector. But, regardless of the number of charged particles encountered, the time delay is always inversely proportional to the square of the transmission frequency. Consequently, the designers of the Transit system found a way to eliminate the transmission time delay. They rigged their satellites to broadcast signals on two different frequencies so that the Transit receivers could mathematically extract out the effect of the ionospheric delay. Two equations in two unknowns provide the desired solution.

Compensating for Tropospheric Delays

Whenever a Transit satellite is situated near the horizon, as seen from the vantage point of the user, its navigation signals pass through a much thicker portion of the earth's atmosphere, thus creating large positioning errors. To minimize the effects of this rather important error source, the user can impose a *mask angle* below which signals from the satellites will not be used for determining longitude and latitude. Whenever a satellite's elevation angle is smaller than the mask angle—typically 5 or 10 degrees—its signals are automatically ignored. This greatly decreases the errors in the navigation solution.

Navigation Techniques

Five or six Transit satellites are typically in operation at any given moment. Depending on longitude and latitude, a typical user can gain access to one of those satellites about every hour or so. For users who are close to the equator—where the satellite orbit planes spread farther apart—the delay between Transit sightings averages about 1.5 hours. For latitudes of 70 degrees or larger (north or south), the user will gain access to one of the satellites, on average, every 30 minutes or so. The actual access intervals are not equally spaced because, over time, orbital perturbations cause the orbits of the Transit satellites to cluster in certain regions of the sky.

When no satellites are visible, the user must employ some alternate (dead reckoning) method of position-fixing, such as inertial or celestial navigation. Then, when another Transit satellite passes into view, the navigation system can be updated to achieve improved positioning accuracy.

Although it has served thousands of navigators extremely well for a number of years, the Transit (SatNav) Navigation System suffers from a number of inherent limitations. In particular:

1. It provides navigation fixes in only two dimensions: longitude and latitude.
2. Its position-fixes are available only about once per hour, the average amount of time required for a new satellite to pass into view.
3. Each time a satellite travels from horizon to horizon the receiver obtains only one position fix. This process typically requires 10 or 15 minutes.
4. During the navigation interval, the receiver must have independent and accurate estimates for its altitude and velocity.
5. All the satellites pass directly overhead at the north and south poles, so position fixes near the poles tend to be rather inaccurate.

These and other serious shortcomings are largely eliminated by the Navstar

Global Positioning System (GPS), which consists of a larger constellation of satellites at much higher altitudes. The Navstar GPS uses pseudo-ranging techniques rather than Doppler shift measurements to fix the user's position. This means that a Navstar receiver measures the signal travel times from several satellites simultaneously, instead of frequency variations from only one satellite at a time.

The Navstar Revolution

The Navstar Global Positioning System satellites are being launched into 10,898–nautical-mile orbits in six orbital planes, each tipped 55 degrees with respect to the equator. The complete constellation consists of 21 Navstar satellites plus 3 active on-orbit spares. The satellites and their ground support equipment are financed by the Department of Defense, but their navigation signals are available free of charge to anyone, anywhere, who cares to use them. Someday the system may serve *millions* of soldiers and civilians.

Navstar Navigation Techniques

A Navstar receiver on the ground, at sea, or in the air picks up the signals from four or more Navstar satellites (either simultaneously or sequentially) to determine its three-dimensional position coordinates: longitude, latitude, and altitude. Figure 1.4 shows how a typical receiver performs the navigation solution, using the signals from four Navstar satellites. A string of precisely timed binary pulses (1s and 0s) travels from the first satellite to the receiver on or near the ground. This takes about one-eleventh of a second. The Navstar receiver estimates the signal travel time by subtracting the time its clock registers from the time indicated by the satellite when it transmitted the relevant pulse. This signal travel time is then multiplied by the speed of light to obtain the estimated range, R_1, to the first satellite.

If the clock in the receiver was perfectly synchronized with respect to the clocks carried onboard the Navstar satellites, three ranging measurements of this type would allow the receiver to determine its three mutually orthogonal position coordinates. However, most Navstar receivers rely on inexpensive quartz crystal oscillators to measure the current time. These crystal oscillators are not synchronized with respect to the much more stable and precise atomic clocks carried onboard the satellites. Consequently, the receiver actually estimates the *pseudo-range* (false range) to each Navstar satellite. All of the pseudo-range measurements are corrupted by the same timing error in the receiver's clock. Thus, the clock bias error (C_B) can be eliminated mathematically by measuring the pseudo-ranges to four satellites instead of only three. This produces a system of four equations in the four unknowns on the right-hand side of Figure 1.4.

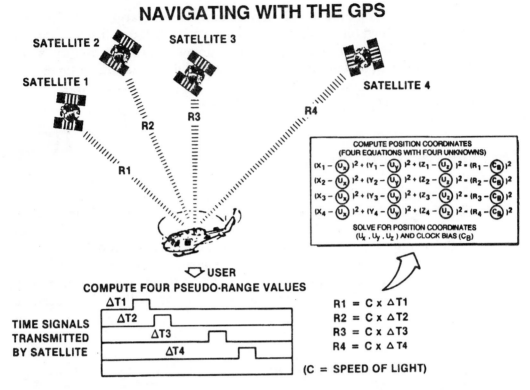

NAVIGATING WITH THE GPS

SATELLITE 2

SATELLITE 3

SATELLITE 1

SATELLITE 4

R4

R3

R2

R1

COMPUTE POSITION COORDINATES
(FOUR EQUATIONS WITH FOUR UNKNOWNS)

$$(X_1 - U_x)^2 + (Y_1 - U_y)^2 + (Z_1 - U_z)^2 = (R_1 - C_B)^2$$
$$(X_2 - U_x)^2 + (Y_2 - U_y)^2 + (Z_2 - U_z)^2 = (R_2 - C_B)^2$$
$$(X_3 - U_x)^2 + (Y_3 - U_y)^2 + (Z_3 - U_z)^2 = (R_3 - C_B)^2$$
$$(X_4 - U_x)^2 + (Y_4 - U_y)^2 + (Z_4 - U_z)^2 = (R_4 - C_B)^2$$

SOLVE FOR POSITION COORDINATES
(U_x, U_y, U_z) AND CLOCK BIAS (C_B)

▽ USER

COMPUTE FOUR PSEUDO-RANGE VALUES

ΔT1

ΔT2

ΔT3

ΔT4

TIME SIGNALS
TRANSMITTED
BY SATELLITE

$R1 = C \times \Delta T1$
$R2 = C \times \Delta T2$
$R3 = C \times \Delta T3$
$R4 = C \times \Delta T4$

(C = SPEED OF LIGHT)

Figure 1.4. If the clock in a Navstar receiver was always precisely synchronized with respect to the atomic clocks onboard the satellites, the timing signals from three satellites would be sufficient to nail down the receiver's three position coordinates: U_x, U_y, and U_z. Synchronization of the receiver's clock, however, is seldom practical, so the timing signals from our satellites are actually required for a three-dimensional positioning solution. The extra timing signal is used to solve for the receiver's clock bias error (c_B), which is common to all four of the reanging solutions and, hence, can be mathematically extracted out.

Three of the circled unknowns are the user position coordinates, U_x, U_y, and U_z. The fourth unknown is the clock bias error, C_B. The three quantities X_1, Y_1, and Z_1 in the first equation are the current position coordinates of satellite number 1. The Navstar receiver determines the three position coordinates of the satellite by picking up the ephemeris constants (orbital elements) being transmitted by satellite number 1. The receiver uses these constants in simple algebraic and trigonometric equations to solve for the satellite's three position coordinates X_1, Y_1, and Z_1 at the time it transmitted the relevant timing pulse.

The Navstar Clocks

Proper operation of the Navstar Global Positioning System requires incredibly precise timing measurements, accurate to within a few billionths of a second. Precise time is so vital because an electromagnetic wave travels one

The Population Explosion in Space

When a Navstar satellite is blasted into orbit, it joins thousands of other man-made objects circling around planet earth. Wander out into your backyard with a good pair of binoculars and you can personally observe the growing population explosion in space. Turn toward the southwest as the sun sinks below the horizon and every few minutes you will be able to spot a satellite glistening in the orange glow of the setting sun.

Wait an hour or two and notice how many stars are visible to the naked eye. If the sky overhead is dark and clear, you should be able to observe about 2,000 stars—roughly the same number as the man-made objects orbiting over a single hemisphere. At this moment NORAD's technicians are tracking more than 6,000 man-made objects orbiting in the vicinity of our beautiful blue planet. They include man-made satellites plus a larger number of spent rockets, clamps and shrouds, and other space-age paraphernalia. Even Ed White's silver glove, which slipped off during his walk in space, is zooming around the earth.

Each year 800 new man-made objects are added to the ones already in space, and roughly half that number plunge back to earth as they gradually lose their drag-battle with the earth's atmosphere. Most of them burn up in the atmosphere, but a few rugged fragments survive their fiery dive back home. Anything smaller than a playground softball is largely invisible to NORAD's radars, but much smaller objects can present a worrysome hazard to unwary travelers in space. Because of its savage velocity, even a fragment as small as a Pinto bean can damage or destroy delicate spacecraft components. NASA experts estimate that the total number of objects of destructive size is at least 15,000, perhaps even more.

Of the trackable objects in NORAD's inventory, 60 percent have resulted from explosions in space, at least 50 of which are known to have occurred. Soviet "killer satellites"—large, deadly sawed-off shotguns—have created one-third of the explosions when they were detonated in low-altitude orbits. Several unplanned explosions of propellant tanks on American upper-stage rockets have also occurred. Some of them were presumed dead in space for many years before they suddenly blew up.

According to *Spaceflight* magazine, astronauts from the former Soviet Union have aggravated the problem of deadly space debris by dumping garbage into space from their manned orbiting platforms. NORAD's radars are not sensitive enough to make a positive identification, but old Russian watermelon rinds may be comingled with today's Navstar satellites and the other orbiting satellites that NORAD's spaceflight experts have been tracking across the purple sky.

foot in one-billionth of a second. Consequently, every one-billionth of a second error in timing translates into at least a one foot navigation error.

The precise timing requirements for the Navstar constellation are met, in part, by installing amazingly accurate atomic clocks in the satellites. The Block II satellites now being launched into space each carry four highly accurate clocks—two cesium atomic clocks and two rubidium atomic clocks. These clocks are so stable and accurate they would lose or gain only about one second every 160,000 years. To help maintain timing precision, fresh clock correction factors are relayed from the ground up to the satellites at least once per day.

When the Navstar constellation was first being developed, an atomic clock of contemporary design was about as big as a household deep freeze. These early clocks were also heavy, power hungry, and extremely temperamental. Fortunately, evolving technology allowed the construction of miniaturized atomic clocks that were small, compact, accurate, and reliable enough to be launched into space. The rubidium atomic clocks onboard the GPS satellites weigh only about 15 pounds, draw 40 watts of power, and maintain timing stability to within 0.2 parts per billion. The cesium atomic clocks, which are slightly more stable, weigh about 30 pounds each.

Practical Benefits for All Mankind

With a fully operational Navstar satellite constellation, no one ever need be lost again. The signals from the satellites—which are available free of charge to anyone, anywhere—have already revolutionized surveying, precision timekeeping, and modern military operations. They are being used in large numbers all over the globe to fix the positions of ships, planes, boats, trains, satellites, even ordinary family Chevrolets.

Navstar receiver costs have dropped dramatically. A decade ago the cheapest available models sold for $140,000 each. Today's least expensive versions are going for less than $2,000 and ordinary hikers and boat owners have been snapping them up by the thousands. Knowledgeable experts maintain that some of tomorrow's models may retail for only $500 each or, perhaps, even less. With costs that low, future applications will be limited only by the dreams and imaginations of those who spend their days—and nights—coming up with clever new ways to use them.

2

The Navstar GPS

The Navstar GPS is a satellite-based radionavigation system that provides continuous global coverage to an unlimited number of users who are equipped with receivers capable of processing the signals being broadcast by the satellites. As Figure 2.1 indicates, the system can be broken into three major pieces or segments:

1. The space segment
2. The user segment
3. The control segment

The fully operational *space segment* will consist of 21 Block II satellites plus 3 active on-orbit spares arranged in six 55-degree orbit planes 10,898 nautical miles above the earth. Each Navstar satellite transmits a precisely timed binary pulse train together with a set of ephemeris constants defining its current orbit.

The *user segment* consists of tens of thousands of Navstar receivers located on the ground, in the air, and aboard ships, together with a few aboard orbiting satellites. A Navstar receiver (user set) picks up the precisely timed signals from four or more satellites—either simultaneously or sequentially—and then computer-processes the results to determine its current position.

If the Navstar satellites could permanently track their precise orbital locations and the exact time, no other hardware elements would be required. Unfortunately, the satellites tend to lose track of where they are and what time it is, so a computer-driven *control segment* is necessary. The control segment includes a group of unmanned monitor stations that track each Navstar satellite as it travels across the sky. This information is then used to determine the satellite's orbital elements, together with any timing errors in its onboard atomic clocks. The resulting corrections are then sent to one of

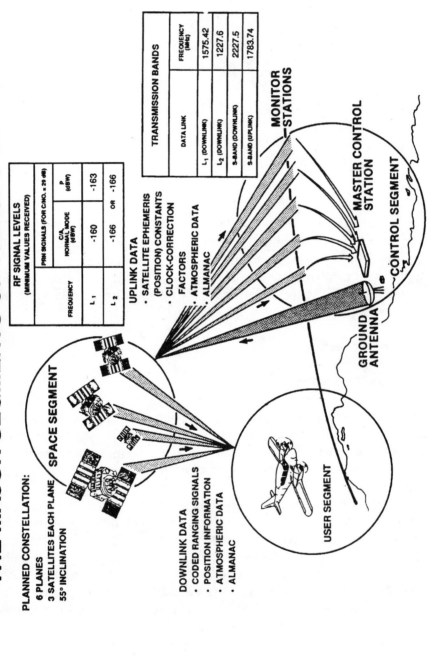

Figure 2.1. The Navstar Global Positioning System can be broken into three major pieces or segments: the *space segment,* which consists of 21 Navstar satellites plus 3 active on-orbit spares, the *user segment,* which consists of tens of thousands of military and civilian receivers, and the *control segment,* which consists of five unmanned monitor stations, a master control station, and a set of ground antennas installed at widely separated locations around the globe.

the four ground antennas that transmit fresh ephemeris coordinates and clock correction factors to the satellites for rebroadcast back down to the users on or near the ground.

The Space Segment

The purpose of the space segment is to furnish accurate timing pulses and satellite ephemeris constants to a worldwide class of users who need to fix their positions, velocities, and/or the exact time. The ephemeris consists of 16 constants that are broadcast to the Navstar receivers so that they can determine where each satellite was when it transmitted its timing pulses. Figure 2.2 depicts the relative locations of the 24 satellites in the Navstar

THE NAVSTAR CONSTELLATION

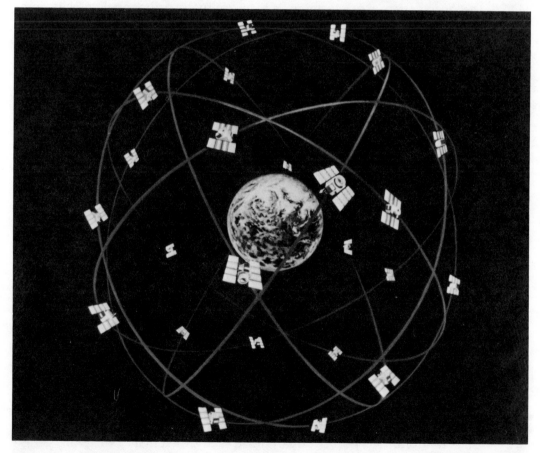

Figure 2.2. The fully operaational GPS constellation consists of 21 Navstar satellites plus 3 active on-orbit spares traveling around 12-hour circular orbits 10,898 nautical miles above the globe. One feedback control loop helps maintain a continous earth-seeking orientation for 12 navigation antenntas on the main body of the spacecraft, and another helps maintain a similar sun-seeking orientation for its two winglike solar arrays.

constellation at a particular instant in time. Notice that, due to perspective, the satellites in the background appear to be smaller than the ones that are closer to the viewer. The winglike solar arrays protruding from the two sides of each spacecraft are always tilted with the proper orientation to catch the perpendicular rays of the sun. The navigation antennas, which are affixed to the lower edge of the spacecraft, always point toward the earth.

Signal Structure and Pseudorandom Codes

To determine its position, a Navstar receiver measures the signal travel times associated with the binary pulse trains from four or more of the satellites. The signal travel time multiplied by the speed of light (186,000 miles per second) equals the slant range from the satellite to the user. By measuring the instantaneous Doppler shift associated with those same four satellites, the receiver can also determine its three mutually orthogonal velocity components.

All of the carrier waves streaming down from the satellites are right-hand circular polarized. This is accomplished by using 12 spiral-wound helical antennas arranged in a tight pattern. Four of them are located in the center quad. The other eight are arranged in a circular ring surrounding the center quad. Circular polarization allows the user-set antennas to access the faint satellite signals without boresight pointing.

Every satellite in the Navstar constellation transmits continuously on the same two L-band frequencies. The reciever uses code-division multiple access to distinguish the satellites from one another. Two different binary codes, the C/A-code and the P-code, are superimposed on the two L-band carrier waves emanating from each satellite. The C/A-code (*Coarse Acquisition Code*) is available free of charge to civilian users all around the world. The P-code (*Precision Code*) is reserved for high-precision military users. It is protected through encryption techniques that restrict access and deny full accuracy to unauthorized users.

Each of the satellites in the Navstar constellation is assigned its own unique C/A-code and its own unique P-code. The C/A-code has a chipping rate of 1 million bits per second, with a repetition interval of 1,023 bits. Thus, it repeats after approximately one one-thousandth of a second. The P-Code has a chipping rate of 10 million bits per second. Its repetition interval is approximately 6×10^{12} bits. Seven days elapse before the P-code sequence repeats. Both the C/A- and the P-codes are pseudorandom binary pulse sequences, with a high degree of "randomness" in their binary 1s and 0s. The "randomness" is only apparent. Actually, the binary pulses are generated by precise mathematical relationships with total predictability.

Phase-shift-key modulation is used to mark the interfaces between the binary 1s and the binary 0s. This means that the L_1 and the L_2 carrier waves experience sharp mirror-image reflections whenever the C/A-code or the P-code switches from a binary 1 to a binary 0 or vice versa. An opportunity for an instantaneous phase shift occurs every one-millionth of a second for the C/A-code and every ten-millionth of a second for the P-code. The L_1 frequency carries both the C/A-code and the P-code transmitted in phase quadrature. This means that they are always 90 degrees out of phase to one another. The L_2 carrier wave is modulated with the military P-code only.

Navigation Solutions

If the Navstar satellites all broadcast on the same two frequencies, how can a Navstar receiver distinguish between the various satellites in the constellation? This is accomplished by code-division multiple access. Each Navstar satellite is assigned its own unique C/A-code and its own unique P-code. Real-time code-matching techniques are used to distinguish among the various satellites and to measure the appropriate signal travel time.

Consider satellite number 1, which is transmitting its own unique C/A-code down toward the users on the ground, as shown in Figure 2.3. This pulse train reaches the ground in approximately one-eleventh of a second or less. The Navstar receiver generates an identical C/A-code pulse train, but it is shifted (displaced) with respect to the pulse train coming down from the satellite.

In order to bring the two identical pulse trains into correspondence, the receiver automatically slews (gradually shifts) the one it is generating. When the two binary pulse trains have been brought into correspondence, binary 1s from the receiver are matched against binary 1s from the satellite, and binary 0s are matched against binary 0s. When coincidence occurs, the *auto-correlation function* suddenly jumps from a value of 0 to a value of 1. This is called "lock-on."

Once lock-on has been successfully achieved, the user set can measure the signal travel time plus or minus the timing error in its quartz crystal oscillator. Fortunately, the magnitude of this so-called "clock-bias error" is the same for each satellite in the constellation. Thus, by measuring the time delays for four or more satellites, the user set can set up a system of four equations in four unknowns to mathematically eliminate the clock-bias error. The four equations used in determining the user-set position are presented at the bottom of Figure 2.3. The four unknowns are U_x, U_y, and U_z (the user's three mutually orthogonal position coordinates) and C_B (the clock bias error in the user-set clock).

These equations cannot be solved explicitly for the four unknown variables, but, of course, they can be solved iteratively using Taylor series expansions. The subscripted variables X_1, Y_1, Z_1 in the first equation are the instantaneous position coordinates of the first Navstar satellite at the time that its timing pulses were transmitted. The user set computes the values of X_1, Y_1, and Z_1 by substituting the ephemeris coordinates streaming down from the first satellite into a set of simple algebraic and trigonometric equations. These precise solutions also require numerical interation.

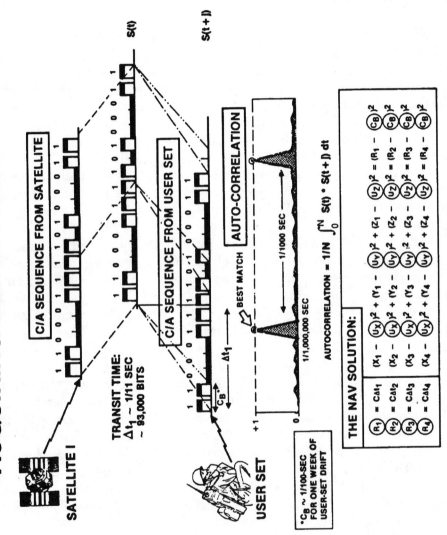

Figure 2.3. Each Navstar satellite transmits two spread-spectrum binary pulse trains, copies of which are created in real time by the user-set electronics. An automatic feedback control loop in the user set skews its pulse train to bring it into correspondence with the identical pulse train being braodcast by the satellite. When correspondence is achieved, the user set can establish the signal travel time plus or minus the clock bias error ≤$_B$. This procedure is repeated for at least three other satellites, to obtain the timing measurements necessary to determine

Navigating Lightning Strikes

Most radionavigation systems determine a navigator's position by measuring the signal travel time of one or more electromagnetic signals. You can use a similar technique to determine the approximate range to an unexpected lightning strike.

First, watch for the flash of lightning. Then count the number of seconds until you hear the sound of thunder. For every five seconds separating lightning flash from thunder sound, there is a one-mile distance between your body and the place where the lightning singed the ground. This simple ranging measurement is based on the fact that the visible light from the lightning flash reaches your eye almost instantly, but the thunder-sound ambles along through the air at only about 1000 feet per second.

How can you refine this primitive "navigation" technique to pinpoint the longitude and latitude of an unexpected lightning strike? One fruitful approach is to send two ground-based observers to widely separated locations. When a lightning strike occurs, each observer counts off the number of seconds until he hears the rumbling sound of thunder. This establishes two circular lines of position, each centered at one of the observer's locations.

In general, the circles will intersect at two different points, one of which corresponds to the location where the lightning bolt touched the ground. Is there any easy way to resolve the solution ambiguity? How about visiting each of the two intersection points to see whether a lightning-damaged tree might be smouldering there?

Thus, we see that the fundamental solution algorithm for the Navstar system is amazingly simple. However, in practice, a number of complicating factors quickly arise. In particular, corrections for three major types of time delays are required:

1. Relativistic time delays
2. Ionospheric distortion
3. Tropospheric distortion

Correcting for Relativistic Time Delays

The relativistic time delays arises in accordance with Einstein's general and special theories of relativity. The clock on board a Navstar satellite ticks at a different rate than the clock in a receiver because the satellites are in a different strength gravitational field and because they are moving at a different velocity than the clock in the Navstar receiver. The pull of the earth's gravity at the GPS altitude is only about 6 percent as strong as the pull of gravity on a receiver on the surface of the earth. The satellite velocity is approximately 12,000 feet per second, compared with 1,000 feet per second or less for the clock in a typical GPS receiver.

Of course, the engineers who design the Navstar system know in advance that, when the satellites are launched into space, they will be in an environment where Einstein's relativistic time-dilation effects will arise with predictable magnitudes. So they instruct the manufacturer to purposely offset the ticking rates of the satellite clocks, to compensate for the relativistic effects they know will later occur.

If the GPS satellites could be launched into perfectly circular orbits, offsetting the ticking rates of their clocks would compensate for virtually all of the relativistic time dilation. Unfortunately, in practice, the orbits of the satellites will always be slightly eccentric. The maximum tolerance in eccentricity for the Navstar orbits is 2 percent. This means that, at their perigee (point of closest approach to the earth), the satellites can be, at most, 288 nautical miles closer to the earth's center than they are at apogee (furthest point from the earth).

As a GPS satellite travels around its elliptical (egg-shaped) orbit, it experiences a sinusoidal variation in relativistic time dilation. As it falls down from apogee toward its perigee, it passes into a stronger gravitational field and its orbital speed increases. This causes the ticking rate of its onboard clock to change with respect to the clock installed in a Navstar receiver on or near the ground. During each 12-hour orbit, the variation in the ticking rate of its clock amounts to, at most, 46 nano seconds (assuming the maximum 2-percent eccentricity).

This time dilation is easily removed by the computer in the Navstar receiver, which substitutes the satellite's known orbital elements into a simple system of closed-form equations. Incidentally, if the Navstar satellites could be launched into perfectly circular orbits 1,800 nautical miles above the earth, the relativistic effects due to the general theory of relativity and the special theory of relativity would nullify one another and no relativistic corrections would be required. However, the earth coverage characteristics at that altitude are not favorable, so a larger constellation would be required.

The relativistic time dilation effects for Navstar navigation are surprisingly large. If the ticking rate of each clock was not offset during manufacture, a typical navigation error after 24 hours without ground updates would amount to approximately 18 nautical miles! The effects due to orbital eccentricity would amount to 100 feet or less.

Correcting for Ionospheric and Tropospheric Delays

When the signals from the Navstar satellites pass through the earth's ionosphere, they are bent and slowed down slightly. The resulting time delay is inversely proportional to the square of the transmission frequency, so two transmission frequencies (L_1 and L_2) can be used to compensate. Each frequency results in a slightly different time delay. Consequently, a simple mathematical correction can be programmed into the sophisticated P-code receivers to extract out nearly all of the ionospheric delay.

The simpler C/A-code receivers pick up only the L_1 signal, so they cannot extract out the ionospheric delay using dual-frequency compensation techniques. Instead, the C/A-code receivers mathematically model the current behavior of the earth's ionosphere, using a set of polynomial coefficients that are included in the satellite's 50-bit-per-second data stream. These coefficients enable the receiver to reduce the ionospheric error by approximately

50 percent, compared with an uncorrected solution. This is usually sufficient for most C/A-code civilian applications.

The tropospheric time delay arises because, when the signals from the Navstar satellite pass through the earth's atmosphere, they are slowed down slightly. The tropospheric delay is quite a bit larger when a satellite is situated near the horizon because its signals pass through a thicker portion of the earth's atmosphere. The time delay is also larger when the Navstar receiver is near the surface of the earth as opposed to being at a higher altitude. The simplest mathematical correction for the tropospheric time delay is a negative exponential function of altitude and a cosecant function of the elevation angle above the local horizon. For demanding applications, more complicated mathematical models are frequently used.

Decoding the 50-Bit-Per-Second Data Stream

A 50-bit-per-second data stream (see Figure 2.4) is superimposed on the C/A-code and the P-code pulse trains coming down from the Navstar

THE CONTENT OF THE GPS DATA STREAM

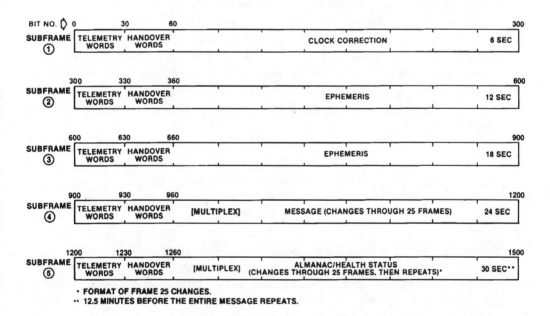

· FORMAT OF FRAME 25 CHANGES.
·· 12.5 MINUTES BEFORE THE ENTIRE MESSAGE REPEATS.

Figure 2.4. The 50-bit-per-second data stream transmitted by each Navstar satellite is superimposed on the pseudorandom C/A- and P-code pulse trains by means of modulo 2 addition. The 30-second frames are further subdivided into five subframes, each of which lasts 6 seconds and contains 300 bits of information. The clock correction factors and the ephemeris constants located in subframes 1 and 2 help the receiver measure the pseudo-range to each satellite so that the receiver can accurately establish its three-dimensional postion coordinates.

satellites. This 50-bit-per-second data stream includes the latest information on the behavior of the satellite clocks and the latest values for their orbital elements (ephemeris constants). The data stream also includes special messages defining the current signal strengths and other health status information for the satellites. This information is frequently updated by the control segment on the ground.

The 50-bit-per-second data stream is divided into 30-second frames, each of which contains 1,500 bits of information. Each frame is divided into five 6-second subframes, each containing 300 bits. Subframe 1 contains the four clock-correction factors, defining the current errors and error growth rates for the satellite clock. Subframes 2 and 3 contain ephemeris constants, defining the current orbit of the satellite.

Subframe 4 contains navigation messages and satellite health status information. Subframe 5 contains the constellation almanac. The almanac is similar to the ephemeris, but there are certain key differences. The almanac contains fewer constants and, consequently, defines a less accurate orbit than the ephemeris. Moreover, the almanac being broadcast by each Navstar satellite defines the orbits to all of the operational satellites in the Navstar constellation.

In each successive 30-second frame the almanac for a different satellite is contained in Subframe 5. Thus, over a transmission interval of about 12 minutes, each satellite tells the users where all the other satellites in the constellation are located in space. The almanac information is not used directly for pseudo-ranging navigation. Instead, it is used in determining which satellites are currently above the horizon and which four are most favorably located to provide the smallest error in the navigation solution.

The 50-bit-per-second data stream is superimposed on the C/A- and the P-code pulse trains using modulo-2 addition. In the modulo-2 addition table:

$$0 + 0 = 0$$
$$1 + 0 = 1$$
$$0 + 1 = 1$$
$$1 + 1 = 0$$

Thus, whenever a binary 1 occurs in the 50-bit-per-second data stream, the modulo-2 addition "inverts" 20,000 adjacent binary digits in the 1 million–bit-per-second C/A-code. Binary 1s become binary 0s and vice versa. A binary 0 in the 50-bit-per-second data stream leaves 20,000 adjacent C/A-code bits uninverted (the same as they were before).

The Various Families of Navstar Satellites

The Navstar satellites can be divided into major families having similar physical characteristics. The initial Block I satellites produced by Rockwell

International each weighed approximately 960 pounds and generated 420 watts end-of-life electrical power. With one exception, all 11 of the Block I satellites reached orbit successfully and all of them operated approximately 3 years or more. Some were still functioning 13 years after launch. The Atlas F booster that was carrying Navstar 7 into space exploded over Vandenberg Air Force Base shortly after liftoff. Each Block I satellite is constructed from 35,000 separate parts. But, when its booster explodes, an even larger number of parts are created. Consequently, no attempt is being made to reassemble Navstar 7.

Navstar 12 was a Block II qualification unit, never intended for flight. The 28 Block II satellites, also produced by Rockwell International, weigh approximately 2,000 pounds each and generate 700 watts end-of-life electrical power. The Block II satellites were originally designed for launch aboard the Space Shuttle, but when the Shuttle Challenger disaster occurred in 1988, they were reassigned to the Delta II expendable booster (medium launch vehicle). All of the Block II satellites launched so far have reached orbit successfully. So far, none have failed.

The 20 Block IIR (replenishment) satellites weigh about 2,300 pounds and generate 1,000 watts end-of-life electrical power. They are produced by General Electric. The Block IIR satellites are scheduled for launch in the mid-1990s.

The User Segment

The purpose of the user segment is to process the time and position information from four or more satellites (either simultaneously or sequentially), to obtain accurate position, velocity, and timing measurements. A Navstar receiver can be divided into three major components: the antenna with its associated electronics, the receiver-processor unit, which picks up the satellite signals and performs the navigation solution, and the control-display unit, which provides information display and a convenient interface between the user and the Navstar system.

A Typical High-Performance 5-Channel Receiver

Multichannel receivers such as the one sketched in Figure 2.5 often serve high-dynamics users, such as military jets. Each separate channel is a radio receiver (front end) that locks on to a separate satellite and tracks it continuously to obtain its instantaneous pseudo-range and its instantaneous Doppler shift. A military avionics receiver of this type often employs two antennas. One is mounted on the top of the aircraft, the other is mounted on its belly. Thus, the receiver can achieve continuous signal reception, even when the aircraft is flying upside down.

A TYPICAL FIVE-CHANNEL USER SET

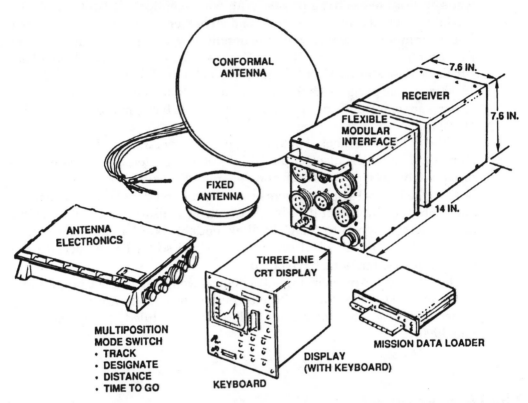

Figure 2.5. Some high-performance military receivers pick up navigation signals from two different antennas mounted on the top and the bottom of the aircraft fuselage. This allows the receiver to gain access to the navigation signals, even when the aircraft is flying upside down. Signals from the active antenna are routed through the antenna electronics to the receiver, which handles the navigation solution. The receiver then feeds the results through the flexible modular interface to the control display unit, where it is presented on the screen in a convenient pictorial or alphanumeric format.

After signal conditioning and amplification, the modulated signals from the appropriate antenna are routed into the receiver unit, which uses computer processing techniques to carry out the navigation solution. This information is, in turn, fed to the control display unit, which usually includes either a liquid crystal display or a cathode ray tube (CRT) screen with a simple keyboard and an (optional) multiposition switch. The control display unit allows the operator to feed in information and to operate the unit in different navigational modes.

The three-dimensional position coordinates, provided by a military P-code receiver, are required to be accurate to within 50 feet SEP (spherical error probable). Such a receiver also provides three mutually orthogonal velocity components with an average accuracy of about 0.3 feet per second.

The velocity components U_x, U_y, and U_z are determined by measuring the instantaneous Doppler shift.

The navigation accuracy available to a particular Navstar user depends on two separate factors:

1. The average User-Equivalent Range Error (UERE) along the line-of-sight vector connecting the user to each satellite.
2. The instantaneous Geometrical Dilution of Precision (GDOP), which defines the geometry of the best four satellites as seen from the user's position on or near the earth.

The 1σ navigation error, which is typically 50 feet or less, is approximately equal to the product of these two quantities (UERE times GDOP).

The largest error component in the User-Equivalent Range Error is caused by clock instabilities and unmodeled space perturbations, which are dominated by the uncertainties in the solar radiation pressure pushing on the spacecraft. Other smaller errors include modeling uncertainties in the satellite ephemeris coordinates and small unmodeled time delays experienced by the electromagnetic signals as they travel through the ionosphere and troposphere. Surprisingly, the overall User-Equivalent Range Error arising from the statistical combination (root-sum-square) of all these different error sources amounts to only about 18 feet when daily control segment updates are routinely achieved. The global time average for the Navstar's GDOP is about 2.3 for a 24-satellite GPS constellation. The product of the average UERE times the GDOP yields an average three-dimensional navigation error for the high-precision military users of only about 50 feet.

The Navstar navigation system provides even smaller errors when it is used for differential navigation. In the differential navigation mode two different receivers pick up the signals from the same four Navstar satellites and exchange information relating to their current navigation solutions. Many of the error components are common to the two solutions, and, consequently, the resulting positioning error is considerably smaller. Depending on the method of implementation, differential navigation errors as small as 6 feet or even less are often achieved.

Operating Procedures

When a Navstar receiver is installed aboard a high-performance military aircraft, its displays are similar to the ones provided by inertial navigation systems, which military aviators have been using for many years. The navigator can typically enter up to 200 waypoints. A waypoint is an intermediate longitude-latitude combination that the airplane must fly through in order to reach its final destination. High-performance military jets are also equipped to process and display moving waypoints and stationary mark points. A *moving waypoint* enables the pilot to rendezvous with another aircraft at or near a specific location. A *markpoint* is a longitude-latitude combination that the pilot needs to mark for future reference. At the push of a button the markpoint is stored in the user set's electronic memory. On a future mission it can be converted into a waypoint to vector the aircraft to the location the pilot marked.

Generally speaking, whenever a GPS receiver is turned on, it automatically provides an accurate navigation solution without manual inputs or human intervention. When it is turned off, it stores its last position coordinates in a nonvolatile electronic memory. When it is turned on again, these coordinates become its estimated position. The nonvolatile memory also stores the last set of almanac constants, defining the locations of all the functioning satellites in the current constellation. These constants are used in the new navigation solution to determine which satellites are above the horizon and which four are the most favorably located to provide an accurate position fix. Even when the user set is turned off, its quartz crystal clock continues to operate. That clock provides the necessary time estimate when the set is later reactivated to obtain a new navigation solution.

The Control Segment

The purpose of the control segment (see Figure 2.6) is to track the GPS satellites and provide them with periodic updates, correcting their ephemeris constants and their clock-bias errors. When the Navstar system was being designed, some of the project engineers proposed tracking the satellites with ground-based lasers reflected from satellite-mounted corner-cube reflectors. However, since the satellites are always transmitting information relating to their current locations, ground-based facilities can be used to *invert* the navigation solution to obtain the desired satellite positioning information.

Inverting the Navigation Solution

A Navstar navigation solution is normally obtained by picking up the C/A- or P-code signals from four or more satellites scattered across the sky. The control segment, in effect, turns the navigation solution upside down. This is accomplished by installing a set of unmanned monitor stations at widely separated locations on the ground. Four monitor stations pick up the navigation signals from a particular satellite at the same time. The monitor stations are independently surveyed to fix their positions, and they are equipped with synchronized cesium atomic clocks to nail down the time to a high degree of precision. The four pseudo-range measurements are then used in an *inverted* navigation solution to fix the location of the satellite and to determine the timing errors in its onboard atomic clock.

THE CONTROL SEGMENT

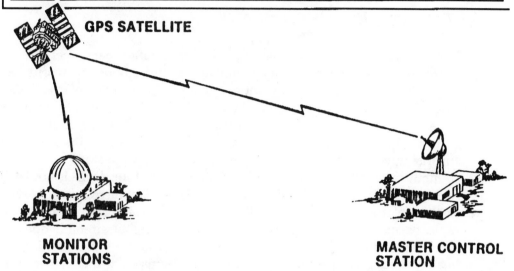

ACCURATELY TRACKS THE GPS SATELLITES AND PROVIDES THEM WITH PERIODIC UPDATES CORRECTING THEIR EPHEMERIS COORDINATES AND THEIR CLOCK BIAS FACTORS

GPS SATELLITE

MONITOR
STATIONS

MASTER CONTROL
STATION

Figure 2.6. Unmanned monitor stations at widely separated locations pick up the signals from each Navstar satellite as it sweeps across the sky. The resulting pseudo-range measurements are then transmitted to the master control station, where they are computer-processed to establish the satellite's clock correction factors and its current orbital elements. These parameters are transmitted to each satellite at least once per day from the four ground antennas positioned around the globe.

In actual practice, hundreds of extra pseudo-range measurements are obtained and used in an overdetermined least-squares solution to improve the accuracy of satellite's ephemeris constants. All of the pseudo-range measurements to each satellite are recorded on wideband tape recordings. These measurements are later transmitted to the master control station, a computer processing facility operated by the U.S. Air Force called the Consolidated Space Operations Center (CSOC). CSOC, which services many military satellites, is located in a secure enclave in Colorado.

The master control station computer-processes the complete collection of pseudo-range measurements to determine the ephemeris coordinates of each satellite and the errors in its onboard atomic clock. This information is then relayed to each satellite once per day on S-band from various 16-foot ground antennas positioned around the globe.

The Monitor Stations and The Master Control Station

Five unmanned monitor stations serve the GPS Block II constellation. They are located at Hawaii, Ascension Island, Diego Garcia, Kwajalein, and Colorado Springs, Colorado. The Master Control Station is located at CSOC in Colorado Springs. The ground antennas used in uploading the satellites are located at Ascension Island, Diego Cargia, and Kwajalein, with a backup shared for other purposes at Colorado Springs.

Field Test Results

Hundreds of realistic field tests conducted by military personnel at Yuma, Arizona, and at other military installations throughout the world have helped to reveal the actual operating characteristics of the Navstar Navigation System. Signal strength and navigation accuracy tests, for example, were vitally important to early mission planners. Ever since they were first installed, the Navstar satellites have been blanketing the earth with surprisingly strong navigation signals. Generally speaking, the signal strength received on the ground has been three to four decibels above the required military specifications. At any given moment, each Navstar satellite blankets 42 percent of the globe with nearly constant-intensity radio frequency transmissions. The maximum signal strength does not occur when a satellite is directly overhead, but rather when it is approximately 40 degrees above the horizon. This stems from the shaped-beam characteristics of the RF signals that originate from the satellite's 12-element helical antenna array.

Early static positioning tests using signals from the Navstar satellites demonstrated highly accurate positioning solutions. Military performance specifications called for a 50-foot average navigation error, but even the early tests provided root-mean-square navigation errors of approximately 24 feet. More recent military tests have yielded similar results even though, on the average, the ground updates have usually been at least several hours older when more recent tests were conducted.

Aerial rendezvous tests using Navstar navigation signals have also demonstrated surprisingly accurate results. In six independent rendezvous tests between an F-4 aircraft and a C-141 military tanker, for instance, the maximum miss-distance between the centerlines of the two airplanes was only about 100 feet. Thus, the biggest centerline miss-distance was only slightly larger than the wing span of the C-141.

Early landing approach tests also provided test engineers with some rather encouraging results. In six independent landing approach tests the C-141 remained well within the boundaries of the instrumented landing system in both azimuth and elevation. The Navstar system thus provided accuracies comparable to existing ground-based landing systems used for landing approach operations.

In early harbor navigation tests in which a fast frigate sailed under the Del Coronado Bridge in the San Diego Harbor, the navigation officers found that some of the buoys marking the boundaries of the safe harbor channel were displaced from their intended locations by as much as 100 feet. Simulated fog increased the difficulty of this early navigation test, but nevertheless the departure trajectory of the fast frigate closely approximated the trajectory it would have followed with ideal viewing conditions under visual navigation control.

3

Performance Comparisons for Today's Radionavigation Systems

Simple radionavigation systems first began to flourish high above the smoke-powder killing fields of World War II. During that conflict, digital pulse trains and serpentine carrier waves guided both Allied and Axis bombers over land and sea. Later, when peace descended on the European continent, derivative ground-based navigation systems helped foster many useful civilian and military applications.

In the late 1950s, when instrument-laden payloads began popping up into outer space, satellite-based transmitters supplemented and replaced the big, spider-beam antennas poking up from the ground. In this chapter we will review the salient characteristics of several representative ground-based and space-based radionavigation systems, and we will compare them with one another on the basis of coverage areas, accuracy levels, transmission frequencies, and the like.

A Sampling of Today's Ground-based Navigation Systems

Far from the rumbling battlefields of World War II, both German and British scientists were highly motivated to develop simple but effective radionavigation systems to help improve the accuracy of bomb delivery techniques. Some World War II systems, such as Decca, are still being used today, although modern refinements have made them more practical and accurate. Others are based on the fundamental architectures of these early World War II systems.

Loran C/D

Loran was one of the earliest and most successful systems for ground-based radionavigation.[1] Two versions are currently in operation: Loran C, which serves civilian users, and Loran D, which serves the military. Often, the transmitters for the two versions of Loran are co-located, especially in the United States.

In its original design, Loran employed time-difference-of-arrival techniques, with the ground-based transmitters working together in pairs to position the user along a hyperbolic line of position on the surface of the earth. The two transmitters, which are today referred to as the *primary* and the *secondary*, are usually separated by a few hundred miles. The secondary transmitter monitors the modulated signals being broadcast by the primary transmitter in order to maintain mutual time synchronization.

A Loran receiver measures the time-difference-of-arrival of the two signals to fix its location on a hyperbolic line of position. By performing similar measurements using the signals from two other Loran transmitters, the receiver can fix its longitude and latitude on the spherical earth.

Loran's carrier waves are transmitted at approximately 100 kilohertz, with wavelengths that are around 1.6 miles long. Thus, the navigation solutions provided by Loran are not particularly accurate. Its ground waves provide coverage over a range of about 1,800 miles with a two-dimensional error amounting to 1,500 feet (at the 95-percent probability level). Sky-wave navigation extends the coverage regime to 2,500 miles, but the signals from the transmitters then reflect off the ionosphere and the positioning error increases to 6,000 to 12,000 feet.

Chains of Loran transmitters cover most of the heavily traveled coastal areas in the Northern Hemisphere. More than 50 transmitters are currently in operation, but all together they provide coverage for only about 10 percent of the globe. At an estimated cost of $60 million, a new mid-continent chain of Loran transmitters is being installed in the midwestern United States. These new transmitters, together with their predecessors, will provide essentially continuous coverage from coast to coast in the lower 48 states.

Omega

Like Loran, Omega was originally designed as a hyperbolic radionavigation system, but it employs phase-difference-of-arrival rather than time-difference-of-arrival techniques to fix the user's position. The Omega system includes only eight very-low-frequency transmitters, but it provides essentially global coverage. On the average, the Omega transmissions cover 88 percent of the globe by day, 98 percent by night. Broader coverage is

---

[1]Loran stands for Long Range Area Navigation

achieved on the nighttime side of the earth because the ionosphere rises in height when nightfall comes and, hence, Omega's reflected signals span a broader coverage area. Although Omega achieves broad area coverage with only eight transmitters, its 2σ navigation error is quite large: 2 to 4 nautical miles. The transmitters operate in the 10 to 13 kilohertz range, with carrier waves that are approximately 16 miles long.

When it is operated in the *differential navigation mode*, Omega's accuracy improves to 1,000 to 2,000 feet. In the differential navigation mode, two or more radionavigation transmitters exchange information with one another, so many of the positioning errors common to the two solutions can be eliminated. In this case, of course, its positioning solutions are measured with respect to the differential base station.

VOR/DME Tacan

VOR/DME Tacan is sued in vectoring civil and military airplanes from one navigation beacon to another along heavily traveled airline routes.[2] The operational characteristics of the VOR portion of the system are sketched in‘ the upper left-hand corner of Figure 3.1. Notice that it employs two different types of signals working together in partnership: a narrow-beam rotating "lighthouse" transmission coupled with a "blinking" omnidirectional pulse. The lighthouse beam rotates at a constant rate of 30 revolutions per second as it systematically circles around the points of the compass. When the rotating beam is pointing toward the North Pole, the blinking signal is briefly broadcast in all directions. Thus, an airplane that is vectoring itself straight toward the transmitter always experiences the same time delay between the receipt of the omnidirectional pulse and the rotating narrow-beam signal. If, for instance, the approach is along a 240-degree azimuth, the airplane will always observe a time delay of one forty-fifth of a second between the receipt of the two different kinds of signals.

The DME portion of the VOR/DME system uses two-way active spherical ranging to measure the slant range between an aircraft and the transmitting station. The avionics system on board the aircraft transmits an interrogation signal toward the DME station, which it immediately rebroadcasts on a different frequency. The slant range to the transmitting station is then obtained by multiplying half the total signal travel time by the speed of light. The operation of the military Tacan portion of this system is similar to the civilian VOR/DME, but it achieves improved accuracies by transmitting at higher pulse rates with higher frequency carrier waves.

The 2σ navigation error for VOR/DME Tacan is approximately 200 to 600 feet. Note, however, that this is actually an angle-measuring system, with an average pointing error of approximately 3 degrees. Consequently, as an

---

[2]VOR = VHF Omnidirectional Range; DME = Distance Measurement Equipment.

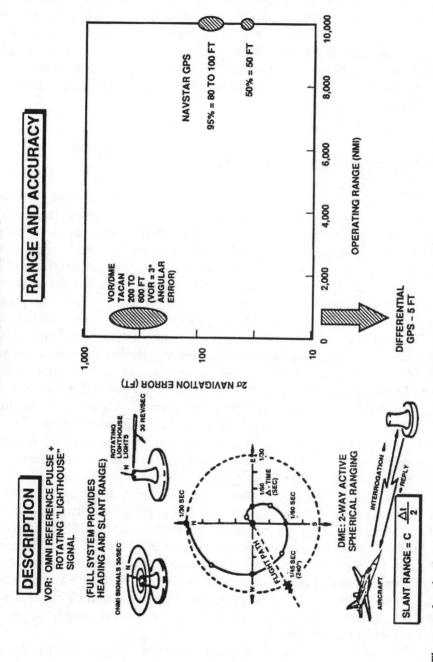

Figure 3.1 Thousands of pilots and navigators vector airplanes, large and small, from waypoint to waypoint using the signals from VOR/DME. The receiver on an airplane determines the proper bearing to a particular transmitter by noting the difference in the arrival times between the blinking omnidirectional signal and the rapidly rotating narrow-beam lighthouse signal. The instantaneous slant-range to the transmitter is determined separately by two-way active spherical ranging.

airplane gets closer and closer to the transmitter, its positioning error systematically decreases.

The Federal Aviation Administration and The Department of Defense operate more than 900 domestic VOR/DME Tacan transmitters, with skimpier coverage being provided throughout the rest of the world. Transmission frequencies range from 108 to 118 megahertz for the VOR portion of the system, and 960 to 1,215 megahertz for the Tacan portion. The corresponding carrier waves are only a few feet long, so, over its relatively short operating ranges, the system can be surprisingly accurate. However, its carrier waves punch through the earth's ionosphere, so only local line-of-sight coverage is achieved.

The Microwave Landing System

The Microwave Landing System (MLS) is used in landing properly equipped airplanes and helicopters at local airports throughout the Western world. The MLS is gradually replacing the ILS (Instrumented Landing System), a simpler radionavigation system that provides airplanes with a localizer pattern (fixed azimuth angle) and a glide-slope pattern (fixed angle of descent). The ILS localizer pattern aligns the airplane's flight path with the airport runway. The glide-slope pattern provides it with the proper sloping line of angular descent. "Marker beacons" at three discrete distances from the end of the runway pinpoint the location of the aircraft with respect to its intended touchdown point. The accuracy of the Instrumented Landing System is approximately 15 to 30 feet. However, ILS is an angular system with an average angular error of 0.05 degrees so that, when the airplane gets closer to its touchdown point, its positioning error systematically declines.

The Microwave Landing System (MLS) uses "windshield wiper" transmissions to help direct aircraft landing operations. One of the two modulated beams sweeps back and forth horizontally to provide the airplane with the proper elevation angle for its gradual descent.

The Microwave Landing System provides roughly the same accuracy as the Instrumented Landing System, but its unique signal structure and its cleverly designed beam manipulation techniques allow pilots to approach the runway along curved or segmented trajectories. This additional flexibility is useful for noise abatement and for landing operations involving helicopters and short-takeoff-and-landing aircraft. NASA's Shuttle Orbiter also uses the Microwave Landing System at Cape Kennedy, Florida, and at Edwards Air Force Base in California, when it touches down on the ground like a butterfly with sore feet.

The Microwave Landing System operates over a range of 17 to 35 miles. Approximately half of all the airplanes licensed for operation in the United States are equipped to pick up the signals being broadcast by the Instrumented Landing System. More than 700 ILS facilities are currently in operation at American airports they service a total of 120,000 users, mostly small private planes. The more advanced MLS system is being developed jointly by the Department of Transportation, the Department of Defense, and the National Aeronautics and Space Administration. Present plans call for the installation of MLS transmitting equipment at 1,250 American airports, but current installations have been falling behind schedule.

The carrier waves for the Instrumented Landing System are centered at 110 and 330 megahertz, with wavelengths that are only a few feet long. The Microwave Landing System operates at 1 and 5 gigahertz, with unusually short wavelengths, each spanning only a fraction of an inch.

Inertial Navigation

Inertial navigation is not a radionavigation system, but it is discussed in the following paragraphs because it is so competitive with the GPS and other radionavigation systems and because it is so popular with so many different kinds of users, especially in the military. Two popular types of inertial navigation systems are in common use: mechanically gimballing (stable platform) systems and strapdown inertial navigation systems.

In a mechanically gimballing system, three mutually orthogonal integrating accelerometers are mounted on a swiveling platform that maintains a constant orientation in inertial space.[3] In a strapdown system the integrating accelerometers are mounted parallel to the body axis of the parent craft. Both types of inertial navigation systems employ gyroscopes to sense altitude changes and integrating accelerometers to measure the three mutually orthogonal acceleration components. The system numerically integrates the acceleration components in real time to establish the three independent velocity components. A second integration provides the three current position coordinates of the moving craft.

An inertial navigation system is a small, self-contained device with a variety of civil and military applications. However, inertial navigation is a dead reckoning technique, so it suffers from one serious limitation: drift-rate errors constantly accumulate with the passage of time. A moderately accurate inertial navigation system will build up a position uncertainty at a rate of about one mile per hour. An extremely accurate system might require ten hours to build up a similar error.

Because its drift errors relentlessly accumulate, an inertial navigation system that operates for an appreciable length of time must be updated periodically with fresh positioning information. This can be accomplished by using an external navigation reference, such as an onboard star tracker or a radionavigation system, such as Loran C or the GPS.

---

[3]Many practical gimballing systems swivel gently to maintain a local vertical orientation. Such a system is based on the Schuler pendulum, a theoretical construct conceptually equivalent to a simple pendulum consisting of a compact mass suspended by a 4,000-mile string. Regardless of how a Schuler pendulum is accelerated, it always maintains a continuous local vertical orientation.

JTIDS Relnav and PLRS

JTIDS Relnav and PLRS are "portable" military communication/navigation systems designed for use in local battlefield areas. The JTIDS (Joint Tactical Information Distribution System) is a ground-based time-division-access communication system coupled with a radionavigation system being financed by the U.S. Air Force. The navigation portion of JTIDS, which automatically reports the user's position, employs active and passive spherical ranging techniques to establish both the position coordinates and the velocity components of the user. This is accomplished by carefully prepositioning several "portable" radio antennas and transmitters in the local battlefield area. The navigation accuracy of JTIDS Relnav is approximately 200 to 300 feet.

JTIDS Relnav is a high-frequency line-of-sight radionavigation system. To extend its primary coverage area, its designers have rigged it so that some of the users who are within the line of sight of the ground-based transmitters can rebroadcast their pulse sequences. This allows users who are situated below the horizon with respect to the ground-based transmitters to achieve fairly accurate position-fixing solutions. JTIDS Relnav uses transmission frequencies in the 960 to 1,215 megahertz range. Its carrier waves are 10 to 12 inches long.

A conceptually similar system, the PLRS (Position Location and Reporting System), is being financed by the U.S. Army. Like the JTIDS, PLRS is primarily a time-division-multiple-access communication system, but it also uses two-way active and passive spherical ranging to provide longitude, latitude, and altitude measurements for large numbers of properly equipped military users. Typical 2σ errors are 200 to 300 feet with respect to the presurveyed transmitters located in the local battlefield area. Transmission frequencies for the PLRS are approximately 420 to 450 megahertz. This corresponds to a sinusoidal carrier wave approximately 3 feet long.

Signpost Navigation Techniques

Signpost navigation employs simple radio transmitters in large numbers to provide reasonably accurate navigation coverage for a local region, such as a construction site or a small city. The automatic vehicle location system operated by the police department in Huntington Beach, California, provides an instructive example of how a signpost navigation system can be implemented. More than 700 navigation transmitters are permanently attached to the wooden and metal telephone poles within the city limits of Huntington Beach. Each telephone pole–mounted transmitter (see Figure 3.2) broadcasts its own unique 11-bit binary code consisting of a prearranged sequence of binary 1s and 0s. That particular pulse sequence identifies the telephone pole from which the transmissions originated.

The 11-bit binary pulse train is picked up by a transceiver located in the

VEHICLE LOCATION METHODOLOGY

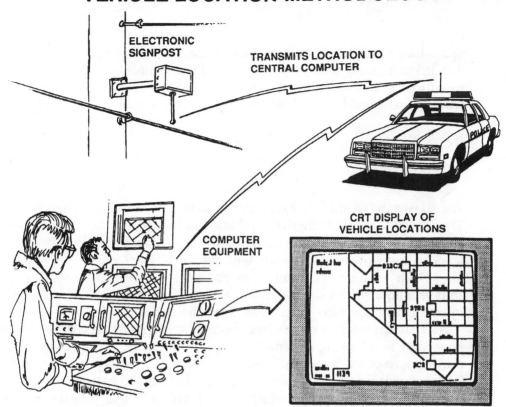

ELECTRONIC SIGNPOST

TRANSMITS LOCATION TO CENTRAL COMPUTER

COMPUTER EQUIPMENT

CRT DISPLAY OF VEHICLE LOCATIONS

Figure 3.2 Positioning measurements for the black-and-white patrol cars in the Southern California city of Huntington Beach are determined by a signpost navigation system with 700 telephone pole–mounted transmitters. Signals from the transmitters are picked up and rebroadcast by a transceiver located in the trunk of each black-and-white patrol car. The signpost navigation system increases the efficiency of patrol car dispatching and helps insure that police officers arrive at the scene of a reported crime in minimum time.

trunk of any nearby black-and-white patrol car cruising the streets of Huntington Beach. The transceiver adds a 7-bit binary code of its own to reveal the identity of the patrol car. It then transmits all 18 bits to police headquarters on Main Street in Huntington Beach, where computers automatically track the movements of the city's fleet of black-and-white patrol cars. When the dispatcher receives a phone call reporting robbery or rape, the address is punched into the computer, which then automatically displays the identities of the three nearest black-and-white patrol cars. The dispatcher then instructs one or more of them to respond to the call.

The positioning accuracy of the Signpost Navigation System at Huntington Beach is determined primarily by the spacing between the city's telephone poles. Typically, the poles are 500 to 2,000 feet apart in the 10-square-mile city, except in the Wildlife Preserve where poles are largely absent, but where serious crimes seldom occur.

A Sampling of Today's Space-based Navigation Systems

Space-based radionavigation systems are extremely expensive to build and install, but, because they are positioned at high altitudes in outer space, they usually provide a much broader coverage area with fewer transmitters than would be required by a ground-based system with comparable capabilities.

Transit

Transit was the world's first successful spaceborne radionavigation system. The Transit satellites were designed and built by engineers at the Applied Physics Laboratory of Johns Hopkins University and at RCA. They were launched into low-altitude polar "bird cage" orbits by the U.S. Navy aboard Thor-Delta boosters. The Transit Navigation System (SatNav) uses Doppler shift techniques to fix the positions of users all around the globe. Its basic operating principles are describe in detail in Chapter 1.

As a Transit satellite travels from horizon to horizon, the user set on the ground picks up its signals and measures the Doppler shift history, which can be represented by a gentle S-shaped curve. The exact contours of the S-shaped curve provide an unambiguous measure of the distance between the satellite's ground track and the user's location on the globe. Users who are closer to the ground trace observe an S-shaped curve that more closely resembles a step function. Those who are farther away from the ground trace observe S-shaped curves with a more gradual curvature.

Five or six Transit satellites are typically providing radionavigation coverage for SatNav users. With a constellation that large, a Transit satellite typically breaches the horizon once every hour or so. It takes the satellite 10 to 15 minutes to travel from horizon to horizon, during which a single position fix is achieved.

The accuracy of the Transit system is approximately 1,500 feet in the horizontal plane at the 2σ level (95-percent probable). Tens of thousands of Transit receivers are in use throughout the world primarily aboard large and small oceangoing vessels and smaller private boats. The dual-frequency navigation signals are broadcast at 150 and 400 megahertz. This corresponds to carrier waves a few feet long.

The Transit system provides global but intermittent navigation coverage to a large and varied group of ground-based consumers who use it with enthusiasm, despite the fact that it suffers from a number of important limitations. In particular, the Transit system provides navigation solutions in only two dimensions, it gives poor accuracy near the poles, and its performance is seriously degraded by any unpredictable motions and any altitude uncertainties during the navigation interval. And, yet, despite these inherent limitations, Transit user sets are extremely popular among navigators throughout the technological world. Tens of thousands of SatNav receivers are currently in use, 98 percent of which are owned and operated by nonmilitary personnel. For several years following the release of the signal specifications to civilian users in 1973, Transit receiver purchases consistently grew at a compound rate of 50 percent per year.

The Navstar Global Positioning System

The Navstar Global Positioning System employs passive spherical ranging techniques to determine the three-dimensional position coordinates for thousands of military and nonmilitary users on or near the earth. A Navstar receiver picks up the binary pulse trains from four or more satellites to measure the pseudo-range to each one. The resulting pseudo-ranges are then substituted into four equations in four unknowns, to solve for the three position coordinates U_x, U_y, and U_z, together with the receiver's clock bias error, c_B.

The Navstar Global Positioning System provides continuous global coverage with a two-dimensional error for the military users of 50 feet at the 50-percent probability level and 80 to 100 feet at the 2σ level (95-percent probable).

Civil users achieve a degraded 2σ error of 330 feet in the horizontal plane.

Saving Race Car Drivers From Outer Space

Serge Goriely was lucky that a specially equipped weather satellite was providing emergency navigation services when his four-wheel drive Citroen whipped out of control in a remote African desert. Otherwise, he would probably not be alive today. Goriely, a 21-year-old professional race car driver, suffered a fractured skull and lay motionless beside his crushed vehicle after it crashed, rolled over several times, and threw him out on the burning sand.

Fortunately, Goriely's car was equipped with a SARSAT search-and-rescue beacon that was automatically activated when he careened off the road. Within seconds, the unit began sending a distress signal through outer space for relay back to Paris. Seventeen minutes later, a doctor was dispatched to the scene of the accident by helicopter, arriving there 79 minutes after the crash. He managed to patch Goriely back together and then admitted him to a nearby hospital for several days recuperation before his colleagues knew for sure that he was to join the 344 other lucky individuals whose lives had been saved by the SARSAT peering down from outer space.

Teams of technicians in the United States, Canada, France, Russia, and seven other technological nations work together to help make sure that SARSAT stays on the air. Emergency beacons—space-age cries for help—stream up to various Russian and American satellites from planes, boats, and even battered race cars for immediate retransmission to rescuers on the ground. The SARSAT units are designed to broadcast coded messages that tell who is in trouble and their approximate location.

Before the SARSAT became available, average notification time for a missing aircraft was 36 to 48 hours. However, if lives are to be saved, rescue efforts must usually be completed within 24 hours or less. With four active SARSATS in a small orbiting constellation, an emergency signal from Africa or any other remote location can be picked up by ground monitoring systems always within one hour after the emergency occurs.

When it is fully operational, the Navstar constellation will consist of 21 satellites plus 3 active on-orbit spares positioned in six orbit planes. Carrier wave frequencies for all the GPS satellites are centered at 1,575.42 and 1,227.6 megahertz for the L_1 and L_2 navigation frequencies, respectively.

The French Argos

The French Argos (see Figure 3.3) is a highly popular and relatively inexpensive space-based radionavigation system. American weather satellites, such as the Tiros relay navigation signals from the Argos transmitters mounted on slowly moving platforms, such as drifting buoys and weather balloons. The Argos system employs bent-pipe radionavigation techniques, whereby the navigation signals follow a sharp angular route from the buoy up to an orbiting satellite and then back down to a special computer processing facility located on the ground. The navigation solution is handled by a dedicated computer processing facility on a non–real-time basis.

Three-dimensional solutions, such as those needed for floating balloons, are typically in error by approximately 10,000 feet. Simpler two-dimensional solutions, such as the ones used in connection with drifting buoys, typically provide 3,000-foot navigation errors. The French Argos is used for a variety of practical applications, including earthquake fault monitoring and the determination of the migration routes of relatively large animals, such as deer and moose.

Side-by-side Performance Comparisons

The transmission frequencies for a sampling of today's most popular radionavigation systems are compared in Figure 3.4. The systems listed on the right-hand side of the figure rely on high-frequency transmissions, so they tend to yield more accurate navigation solutions. These include the Navstar GPS, the Microwave Landing System, the JTIDS relnav, and the Army's PLRS.

The Navigation systems spotted on the left-hand side of the figure, such as Omega and the Loran, rely on lower frequency transmissions, which generally result in relatively inaccurate navigation. The transmission frequencies for Omega and Loran were specifically selected so that their carrier waves would reflect off the charged particles in the earth's ionosphere. This greatly increases the coverage regimes, but at a substantial sacrifice in navigational accuracy.

Comparisons between the anticipated positioning errors and the operating ranges for various popular radionavigation systems are presented in Figure 3.5. The navigation errors all apply to the 2σ level (95-percent probable). Notice that the horizontal scale ends at 11,000 nautical miles (the half-circumference of the earth). When a system's operating range is posi-

THE FRENCH ARGOS SYSTEM

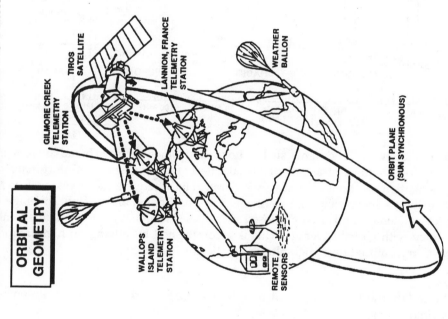

Figure 3.3 The French Argos is a bent-pipe radionavigation system used to track drifting buoys and migrating animals to an accuracy of a few thousand feet. A continuous tone (sinusoidal carrier wave) is relayed from each transmitter up to an orbiting satellite. Then it is sent back down to a ground-based computer processing facility, where the navigation solution is executed.

COMPARISON OF THE VARIOUS
TRANSMISSION FREQUENCIES

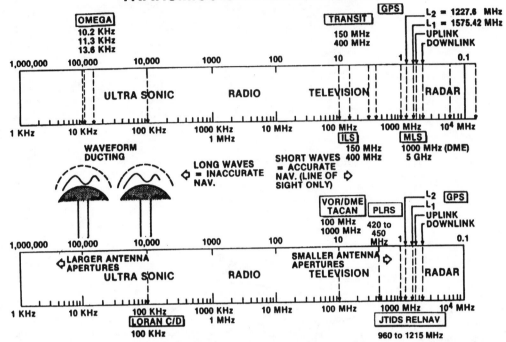

Figure 3.4 Radionavigation systems, such as Loran and Omega achieve broad-area coverage by using very-low-frequency transmissions that reflect off the ionosphere. Unfortunately, the resulting navigation solutions are not very accurate. High-frequency systems, such as MLS and JTIDS Relnav, achieve enhanced accuracies but with limited line-of-sight coverage. The GPS, the Soviet Glonass, and other proposed space-based radionavigation systems combine good accuracy with global coverage. This is accomplished by relying on high-frequency carrier waves transmitted from high-altitude platforms in space.

tioned on the 11,000-nautical-mile line at the far right, it provides essentially global coverage.

Radionavigation systems capable of providing global coverage include the Navstar GPS, the Transit Navigation System, and Omega. At the 2σ probability level, the Navstar GPS yields a positioning error of 80 to 100 feet when it is operated in its undegraded military mode. Unauthorized (civil) users, who are purposely restricted from achieving the full military accuracy of the system, typically experience a pseudo-ranging navigation error of 330 feet. The 2σ error for the Transit Navigation System is about 20 times bigger than the comparable error for the Navstar GPS. The Omega also provides essentially global coverage, but with a much larger positioning error of 2 to 4 nautical miles. Thus, the Omega System is about 100 times less accurate than the Navstar GPS. Some short-range positioning systems, such as ILS and MLS, achieve positioning accuracies that are fully competitive with the Navstar, but only within a small line-of-sight coverage area.

RANGE AND ACCURACY COMPARISONS FOR VARIOUS NAVIGATION SYSTEMS

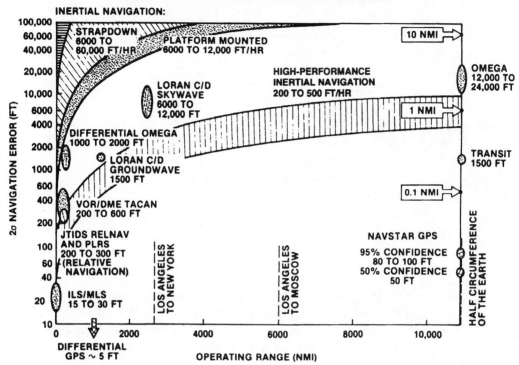

Figure 3.5 Worldwide navigation coverage is provided by the Navstar Global Positioning System, which yields a military accuracy of 80 to 100 feet 95 percent of the time. Other radionavigation systems with global coverage characteristics include the space-based Transit, which is about 20 times less accurate than the GPS, and the ground-based Omega, which is 100 times less accurate. Short-range systems with roughly comparable performance—within their limited coverage regimes—include the Microwave Landing System, which is used in landing specially equipped airplanes, and the JTIDS Relnav, a military system designed for use in local battlefield areas.

The Navstar Global Positioning System provides today's navigators with a number of advantageous characteristics. These include passive ranging with an unlimited number of users, accurate position and velocity measurements, good jamming immunity, precise time synchronization, and high-dynamic operation. The Navstar GPS also provides continuous navigation coverage 24 hours a day, with a fresh navigation solution every second or so. The receivers used in connection with some alternate navigation systems may be a bit cheaper, but no other system can provide comparable services at a comparable price.

4

User-Set Architecture

The electromagnetic signals picked up by a Navstar receiver are surprisingly whispy and tenuous. Their energy density is roughly equivalent to the illumination from the brake light of a Lincoln Town car seen by another driver 1,500 miles away. Stand on the observation deck of the Chicago Sears Tower, looking toward the western sky, and your face will be bathed by a comparable amount of energy from a 15-watt Christmas tree bulb clipped to the Nativity scene in San Francisco on the Pacific Coast.

And yet, despite their ephemeral nature, those faint Navstar signals can be received by surprisingly small antennas, some of which are only two inches across. Once they have been picked up, the Navstar navigation signals can be processed by miniature receivers, some as small, compact, and inconspicuous as a king-size pack of cigarettes.

Many modern receivers are jam-packed with today's most advanced gallium-arsenide computer chips, chips that feature low power consumption, fast processing, and glitch-free operation. Working with highly efficient antennas, advanced architectures, digital processing techniques, and clever software routines, today's electronic engineers are turning the next generation of Navstar receivers into amazingly capable machines.

The Major Components of a Typical Navstar Receiver

Practical and accurate navigation services were successfully provided by the earliest Navstar receivers, but, generally speaking, those early devices were bulky, heavy, expensive, and difficult to operate. Even a reasonably compact model, such as the Texas Instruments 4100, weighed 40 pounds and was roughly the size of a small electric typewriter. The initial version retailed for

$139,000. Later it was reduced to $119,000, before it was replaced by a more competitive design.

A few of today's models are only a bit bigger than pocket calculators some sell for only a few hundred dollars each. But, regardless of size, price, or complexity, a modern receiver can usually be broken down into five major subassemblies:

1. The receiver antenna and its associated electronics
2. The tracking loops
3. Navigation processor
4. Power supply
5. Control display unit

Figure 4.1 highlights various interactions between these major components, and summarizes the critical functions performed by each one of them.

The Receiver Antenna and Its Associated Electronics

A Navstar antenna is designed to pick up the right-hand circular-polarized L_1 and/or L_2 carrier waves from selected satellites located above the horizon. It then concentrates and amplifies the modulated carrier waves, and converts their electromagnetic energy into an equivalent electric current still containing the appropriate C/A-code, P-code, and data stream modulations. Often, the Navstar antenna is physically isolated from multipath reflections by means of a flat metal mount called the "ground plane," which helps prevent smearing and distortion of the binary pulse trains.

The L-band signals picked up by the antenna are routed through a preamplifier circuit that boosts their power level for easier processing by the downstream electronic devices. The signals from the antenna are often routed to the receiver through a coaxial cable which also carries electrical power from the receiver to the antenna's preamplifier circuits.

The Tracking Loops

Two different types of tracking loops are executed by a Navstar receiver: (1) the *code-tracking loop,* which tracks the C/A- and/or P-code pulse trains to obtain the signal travel time for each relevant satellite, and (2) the *phase lock loop,* which tracks the satellite's carrier wave to obtain its instantaneous Doppler shift. Code tracking allows the receiver to measure the appropriate *pseudo-ranges* to the four (or more) satellites necessary for an accurate positioning solution. Doppler shift tracking allows the receiver to measure the corresponding *pseudo-range rates* so it can estimate accurate values for the receiver's three mutually orthogonal velocity components.

A special mixer (multiplier circuit) steps down the frequency of the carrier

GPS RECEIVER FUNCTIONS

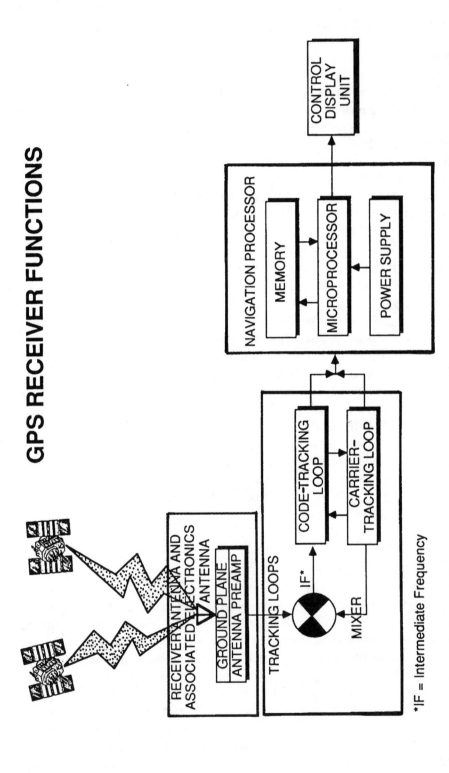

Figure 4.1 This simple block diagram highlights the major component subassemblies of a typical Navstar receiver. The receiver antenna picks up the signals from the satellites, and amplifies them before they are fed into the two tracking loops that lock onto the carrier waves and the appropriate binary codes to obtain pseudo-range and the Doppler shift measurements. Once these measurements have been made, the microprocessor automatically determines the user's current position coordinates and velocity components, which are displayed by the control display unit in a convenient human-oriented format.

waves from gigahertz to megahertz before they enter the tracking loops. Carrier waves at an intermediate frequency are created by mixing the incoming satellite signals with a pure sinusoidal oscillation created by a voltage-controlled crystal oscillator. The intermediate carrier wave still contains all of the C/A-code, P-code, and data stream modulations broadcast by the satellite, but they are all shifted to the intermediate beat frequency, which is much easier for the downstream circuits to process.

The code-tracking loop automatically generates and time shifts a replica of the 1-million bit-per-second C/A-code coming down from each satellite to bring the replica into precise correspondence with the satellite's C/A-code. When matchup (auto-correlation) has been achieved, the auto-correlation function suddenly jumps from a value of 0 to a value of 1.

Auto-correlation (lock-on) allows the code-tracking loop to measure the instantaneous pseudo-range to the satellite and also to decode its 50 bit-per-second data stream. Most P-code receivers use the Handover Word located in each 6-second subframe of the data stream to substantially decrease the time required to lock on to the 10-million bit-per-second P-code.

The *carrier-tracking loop* uses a voltage-controlled crystal oscillator to create a replica of the incoming carrier wave. It then beats the two carrier waves together to determine their beat frequency, an indirect measure of their relative Doppler shift. Carrier-wave tracking is often accomplished by using some variation of the Costas Loop, which was developed and perfected in 1959 by J. P. Costas at General Electric in Schenectady, New York.

Navigation Processor

The navigation processor uses the pseudo-range and the Doppler shift measurements to determine the instantaneous position coordinates and the velocity components of the Navstar receiver. The solid-state processing circuits also handle the satellite selection algorithms, any necessary coordinate transformations, the Kalman filtering techniques, and the routing calculations needed for efficient waypoint navigation.

The memory units in the navigation processor provide erasable "scratch pad" storage for the various types of computations. Each time the receiver is turned off, nonvolatile portions of its microprocessor memory are used to save the last set of position coordinates, together with the last set of almanac constants. When the receiver is turned back on again, these values are used to obtain the "first guess" estimates of position and to determine which four satellites are the most favorably positioned for accurate navigation.

For some specialized applications the microprocessor's memory is used to store large arrays of pseudo-range measurements for precise postprocessing. In postprocessing applications, improved values for the satellite's ephemeris constants obtained after the fact are used to enhance the accuracy of delayed navigation solutions. Surveying and military test range applications, for instance, obtain substantial accuracy improvements by using appropriate post-processing techniques.

Power Supply

The DC power needed to operate a Navstar receiver is usually provided by disposable lead-acid batteries or rechargeable nickle-cadmium (Ni Cd) batteries. But the electrical systems of trucks and tanks can also provide the requisite power. In some cases, the power supply module includes power conversion devices (AC to DC) and the devices needed to condition and convert the recharging power supply.

Control-Display Unit

The control display unit is a convenient man-machine interface between the user and a Navstar receiver. It is designed to accept inputs and instructions from the user, including the desired operating modes, stationary and moving waypoints, coordinate systems, and any necessary encryption keys.

The current position and velocity are automatically displayed on light-emitting diodes (LEDs) or cathode ray tube (video) screens. The control display unit also displays the exact time and waypoint navigation instructions under efficient user control.

Choosing the Proper User-set Architecture

As Figure 4.2 indicates, three different types of Navstar receivers with fundamentally different architectures are currently available in the commercial marketplace:

1. Continuous-tracking receivers
2. Slow-sequencing receivers
3. Fast-sequencing receivers

A *continuous-tracking receiver*, which is also called a *multichannel* receiver, tracks four or more Navstar satellites continuously with a separate front-end being devoted to each satellite being tracked. It gains uninterrupted access to the 50-bit-per-second data stream superimposed on the transmissions being received from each satellite. Continuous-tracking receivers are more expensive than the two alternate architectures, but they are simpler in concept and can operate successfully under high-dynamic, military conditions.

MAJOR TYPES OF RECEIVERS

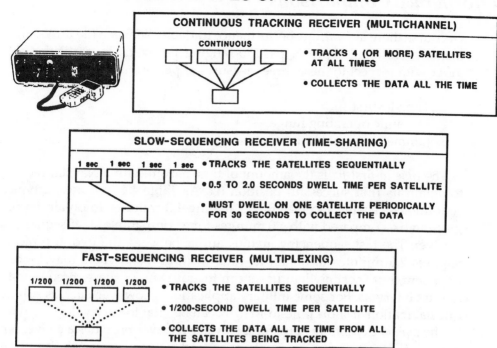

Figure 4.2 Three major types of pseudo-ranging receivers are currently available in the commercial marketplace. Though costly, the *continuous-tracking receiver* is an excellent choice for applications that require a rapid time-to-first-fix, high-dynamic operation, and/or good jamming immunity. The *slow-sequencing receiver* is a cheaper alternative for less demanding applications. The *fast-sequencing receiver* is generally between the two in terms of performance and price.

A *slow-sequencing receiver*, which is also called a *time-sharing receiver*, tracks four or more Navstar satellites sequentially. It typically dwells on each satellite for one or two seconds before moving on to the next satellite in sequence. A slow-sequencing receiver is similar in concept to a time-sharing computer network in that the demands on the system are serviced sequentially using the same hardware. A slow-sequencing receiver typically dwells on each satellite for two seconds or less, so it must interrupt its processing sequence periodically in order to obtain at least one 30-second frame from the satellite's 50-bit-per-second data stream.

A *fast-sequencing receiver*, which is also called a *multiplexing receiver*, tracks four or more Navstar satellites sequentially, but it dwells on each of them for an extremely brief interval. If, for instance, a multiplexing receiver is sequencing between four different satellites, it might dwell on each one for one two-hundredth of a second. Thus, it gets back to each satellite just in time to pick up the next bit in its 50-bit-per-second data stream. In this way a fast-sequencing receiver can gain access to all of the 50-bit-per-second data streams from all four of the satellites at all times.

Performance Comparisons

Of course there are various performance tradeoffs between the three fundamentally different types of pseudo-ranging receivers. In particular, they differ with respect to these three measures of performance:

1. Time-to-first-fix
2. Dynamic operating range
3. Jammer-to-signal ratio

The *time-to-first-fix* is the amount of time required for a Navstar receiver to obtain its first successful position fix. As Table 4.1 indicates, a typical continuous-tracking receiver might require 1.3 minutes to obtain its first position fix, compared with 4.0 minutes for a comparable slow-sequencing receiver. The fast-sequencing architecture is between the two. It typically requires 2.5 minutes to obtain its first position fix. Owners may become impatient, but most civilian users can tolerate any of these three time-to-first-fix intervals. For some military applications, however, such as submarine navigation, a short time-to-first-fix can be vital to survival.

The *dynamic operating range* for a Navstar receiver provides a convenient numerical measure for the acceleration and velocity uncertainties under which it can operate successfully. A well-designed continuous-tracking receiver, for instance, might operate successfully in a 10-g dynamic environment, compared with a 1-g environment for a comparable slow-sequencing receiver. A fast-sequencing receiver of similar design might function well in an intermediate dynamic environment of about 4 g's. Thus, a continuous-tracking receiver is suitable for use aboard high-performance military jets, which, except for emergencies, stay well below 9 g's. A comparable slow-sequencing receiver is suitable for use aboard a ground-based vehicle, such as a tank or an army truck. A fast-sequencing receiver is suitable for use aboard vehicles that experience intermediate dynamics, such as jet-powered boats and crop dusting planes.

The *jammer-to-signal ratio* for a Navstar receiver quantifies the amount

Table 4.1   The performance capabilities of various types of receivers[1]

Type of receiver	Typical time To first fix (min)	Typical dynamic operating range (Gs)	Typical jammer-to-signal ratio (dB)[2]
Continuous tracking	1.3	10	−10
Slow sequencing	4.0	1	−11
Fast sequencing	2.5	4	−16

[1]A variety of engineering assumptions enter into estimates of this type. The values presented in this table are intended to indicate trends, not to characterize the salient performance characteristics associated with receivers of each type currently available for purchase.
[2]Relative values.

of jamming immunity it provides. If a continuous-tracking receiver is baseline at −10 dB, then comparable slow-sequencing and fast-sequencing receivers might typically provide −11 dB and −16 dB jammer-to-signal ratios, respectively. Thus, a slow-sequencing receiver is only slightly more susceptible to jamming than a multi-channel receiver, but is considerably less immune to jamming than a fast-sequencing (multiplexing) receiver. Fast-sequencing architectures are thus unsuitable for use aboard military vehicles that must operate in unfriendly environments populated by enemy jamming devices.

Selecting the Antennas

The L-band transmissions streaming down toward earth from the Navstar satellites arrive with extremely low power densities, so reasonably efficient user-set antennas are needed to pick up the faintly modulated signals they contain. The satellite transmissions are all right-hand circular polarized, so the user-set antennas must also be right-hand circular polarized. However, the antennas need not be bore-sighted toward any particular sector of the sky to acquire the signals from the Navstar satellites. For efficient operation, the user-set antennas must be positioned so that they are free of major obstructions. Thick foliage, for instance, blocks the satellite signals, especially if the foliage is dense enough to stop the light from the sun.

A few years ago a team of researchers at the FAA research center near Atlantic City, New Jersey, conducted a series of tests to determine if the Navstar signals could successfully penetrate a helicopter rotor blade. They found that rotor-blade blockage was not a major problem—unless the antenna was positioned under the hub of the rotor blade where it was blocked by essentially solid metal. Intermittent interruption by the whirling rotor blades did not interfere with accurate Navstar navigation.

Operational Navstar receivers employ antennas in a variety of shapes, including dome antennas, patch antennas, volute (spiral) antennas, and blade antennas. Recently, Ball Aerospace has successfully marketed a square ceramics microstrip antenna only 2 inches on a side (see Figure 4.3). The C/A-code version (L_1 carrier wave only) is 0.1 inches thick and weighs only one ounce. Another version, which is rigged to pick up both the L_1 and the L_2 signals, has the same 2-inch cross section, but it is 0.3 inches thick. It weighs 2.7 ounces. Power-limited and size-limited designs—such as hand-held receivers and those designed for use in connection with automotive and avionics-type applications—are using large numbers of these tiny micro-strip antennas.

A few high-performance military applications call for electronic null-steering antennas. A null-steering antenna is an electronically steered phased array usually constructed from seven antenna elements mounted side by side on a flat ground plane. The entire unit is flush-mounted with the skin of an aircraft. With null-steering techniques the nodes of the antenna

CERAMIC GLOBAL POSITIONING SYSTEM ANTENNAS

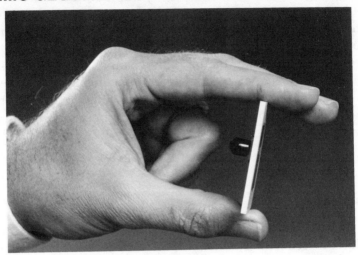

| BALL AEROSPACE GPS ANTENNA |

- PATENTED CERAMICS MICRO-STRIP ANTENNA

- 2" x 2" AND ONLY 0.1" THICK: ONE OUNCE

- THICKER ANTENNA (0.3" THICK WEIGHING 2.7 OUNCES) RECEIVES BOTH L_1 AND L_2 BROADCASTS

Figure 4.3 Ball Aerospace produces two small micro-strip antennas for use in connection with Navstar receivers. Both are 2-inch squares. The single-frequency L_1 version is one-tenth of an inch thick and weighs 1 ounce. The more complicated L_1/L_2 version is three-tenths of an inch thick. It weighs 2.7 ounces. According to Ball Aerospace brochures, these 2-inch antennas match the performance capabilities of the 5-inch antennas that the company produced in 1975.

pattern are automatically directed toward the satellite signals. This increases the average jamming immunity of the unit by a substantial amount, thus opening up highly demanding military applications and making other military applications considerably more successful.

Selecting the Proper Computer Processing Techniques

The Navstar Global Positioning System represents a beautiful marriage between space technology and computer technology. The five-channel military receiver built by Rockwell Collins, for instance, is computationally equivalent to a desk-top computer. It contains five front-end chips, each of which can perform 200,000 floating-point operations per second. A sixth chip capable of performing 400,000 operations per second handles the navigation solution. Thus, a typical military receiver of modern design is capable of performing at least 1,400,000 mathematical operations every second.

The software routines in a modern receiver are also surprisingly extensive. Sixty thousand lines of code are needed for successful operation of a five-channel military receiver. At Rockwell International's Government Avionics Division in Cedar Rapids, Iowa, software development required the services of 65 full-time professional programmers. Modularization, layered architecture, and effective data-hiding techniques were coupled with structured programming and professional software engineering, but developing the necessary software was still a difficult undertaking.

Solving for the User's Position

How does a Navstar receiver obtain a pseudo-range measurement to each of the relevant Navstar satellites? Each satellite repeatedly broadcasts its own unique C/A-code, a prearranged sequence of binary 1s and 0s that identifies it to the receiver, which attempts to lock onto the satellite by generating an identical sequence of binary 1s and 0s. Of course, the two pseudorandom codes will be offset (displaced) with respect to one another. In order to "lock on" to the signal coming down from the satellite, the user set automatically slews (shifts) its sequence of binary 1s and 0s to bring the two sequences into correspondence. When this occurs, the auto-correlation function suddenly jumps from a value of 0 to a value of 1, as shown in Figure 4.4. Once it has locked onto four or more GPS satellites, a C/A-code receiver is ready to perform its first navigation solution.

P-code receivers go through a more elaborate two-step procedure. First the receiver locks onto the C/A-code pulse sequence in the manner just described. This allows it to gain access to a special "Handover Word" in each subframe of the 50-bit-per-second data stream. The Handover Word contains a set of constants that allow the receiver to generate the current P-code, thus greatly shortening the necessary search.

P-code receivers have a number of intrinsic advantages over their simpler C/A-code counterparts. In particular, a P-code receiver provides more accurate navigation, enhanced jamming immunity, and better multipath rejection.[1]

Once a Navstar receiver has successfully locked on to four or more Navstar satellites, it can insert the four measured pseudo-ranges into a system of four equations in four unknowns so that it can solve for its position coordinates U_x, U_y, and U_z and its clock bias error, C_B. By measuring the Doppler shift (compression of the sinusoidal carrier waves) from the same four Navstar satellites, the receiver can also determine its three mutually orthogonal velocity components. Undegraded velocity errors as small as 0.3 feet-per-second (1σ) are relatively easy to achieve.

---

[1]Multipath interference occurs when the signals from a Navstar satellite reflect off adjacent objects, such as a nearby lake or the wings of a plane. Multipath interference smears the binary pulses, thus reducing the accuracy of the pseudo-range measurements. Annoying "ghosts" on a T.V. screen are created in a similar manner.

THE NAVIGATION PROCEDURES:
SOPHISTICATED USERS

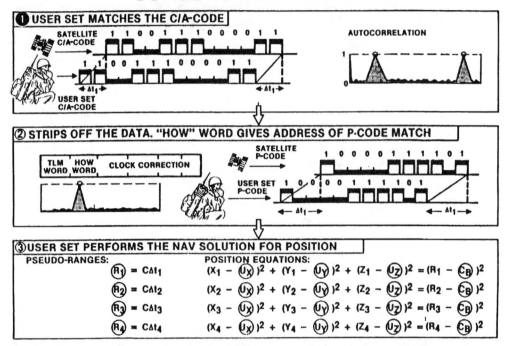

Figure 4.4 Most P-code receivers go through this relatively complicated two-step procedure to lock onto the P-code pulses of each favorably situated satellite. First the receiver generates the appropriate 1-million bit-per-second C/A-code, and then slews (shifts) that code to bring it into correspondence with the identical C/A-code coming down from the satellite. It then decodes the 50-bit-per-second data stream, which contains the Handover Word in each 6-second subframe. The Handover Word helps the receiver generate the satellite's current P-code and match it after only a short search.

Computing and Interpreting the Geometrical Dilution of Precision

The accuracy of a Navstar navigation solution depends on two key factors: (1) the accuracy with which the receiver can measure the slant range to the appropriate satellites, and (2) the geometrical locations of the satellites with respect to one another, as seen from the user's point of view. Unfavorable satellite geometry occurs, for instance, whenever the satellites being tracked happen to lie along a straight line across the sky or whenever they are bunched up together.

To obtain the smallest navigation error, the satellites should be widely dispersed with big angles betwen them, as seen from the user's location. For the Navstar constellation the optimum geometrical pattern (biggest possible tetrahedron) occurs when one of the satellites is directly overhead, and the other three are spaced 120 degrees apart along the horizon. The so-called

Table 4.2 Geometrical Dilution of Precision The Various Types

Type Of GDOP	Interpretation	Coordinates Involved	Typical interested User
GDOP	Geometrical Dilution of Precision[1]	U_x, U_y, U_z, U_t (3-D coordinates plus time)	Mostly of theoretical interest Moving time sync users
PDOP	Position Dilution of Precision	U_x, U_y, U_z (3-D coordinates)	Air-related and space-related users
HDOP	Horizontal Dilution of Precision	U_x, U_y (local horizontal coordinates)	Maritime users
VDOP	Vertical Dilution of Precision	U_z (Altitude)	Air-related users
TDOP	Time Dilution of Precision	U_t (Time)	Time sync users

[1]GDOP (Geometrical Dilution of Precision) is a specific term that refers to navigation precision having to do with $U_x, U_y, U_z,$ and $U_t.$ It is also a *generic* term that refers to *any* of the five different dilutions of precision listed in this table.

Geometrical Dilution of Precision (GDOP) provides a convenient numerical measure of how well the satellites are mutually positioned. The smallest (most favorable GDOP) occurs when the unit tetrahedron has the maximum possible volume. The unit tetrahedron is formed by pointing four unit vectors from the receiver toward the four satellites, and then closing off the tetrahedron that results.

As indicated by Table 4.2, five popular types of GDOPs are in common use; the non generic GDOP is mostly of theoretical interest. It relates to the geometrical error-magnifier in the three position coordinates $U_x, U_y,$ and $U_z,$ combined with the error in time, $U_t.$

The *Position Dilution of Precision* (PDOP) relates to the position uncertainty in the three mutually orthogonal position coordinates $U_x, U_y,$ and $U_z.$ PDOP is important to air-related users. The *Horizontal Dilution of Precision* (HDOP) relates to the two mutually orthogonal position errors in the horizontal plane. HDOP is of interest to maritime users who already have a good handle on their vertical position—since they are usually sitting on the surface of the sea.

The *Vertical Dilution of Precision* (VDOP) relates to the error in the vertical (altitude) component. This measure of performance is important to airplane pilots who are attempting to execute precise landings at non instrumented airfields. *Time Dilution of Precision* (TDOP) relates to the error in the current time. The TDOP value is important to scientists and engineers who are attempting to mutually synchronize distant atomic clocks.

Under optimal conditions, when one satellite is directly overhead and the other three are on the horizon 120 degrees apart, the PDOP equals $\sqrt{8/3}$, or 1.639. To obtain the current three-dimensional position error, we multiply the 1σ ranging error to the satellites by the PDOP.

In practice, a PDOP of 1.639 is seldom achieved, because we usually impose a mask angle below which the navigation signals from the GPS satellites will not be used. The mask angle helps minimize the distortions in the navigation signal that would otherwise occur when the carrier waves travel through the thicker portions of the atmosphere down near the horizon. Depending on the application, mask angles typically range between 5 and 15 degrees.

Ranging Error Budgets

The observed ranging errors to the Navstar satellites can be estimated by constructing a user-equivalent range error budget that lists the values of the various independent error components. A typical sample for P-code navigation is represented by the bar charts at the top of Figure 4.5. Notice that the components are the satellite clock and ephemeris errors that, taken together, amount to 12.8 feet. Other errors include the unmodeled iono-

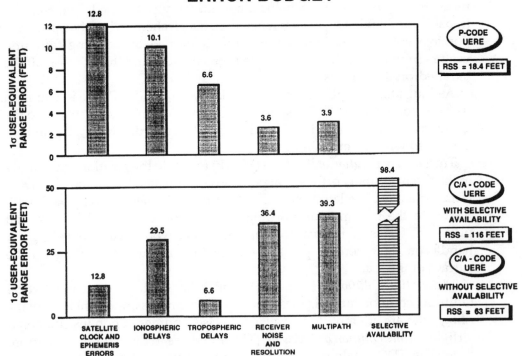

Figure 4.5   The two rows of bar charts in this figure represent error components arising from the measurement of the signal travel time between the user and a typical GPS satellite. When the various error components are combined, the P-code users end up with a total 1σ ranging error of 18.3 feet. The corresponding 1σ errors for C/A-code users with and without selective availability are 116 and 63 feet, respectively.

spheric and tropospheric delays, receiver noise and resolution, and the multipath error. An estimate of the total combined ranging error can be obtained by assuming that the various error components are statistically uncorrelated. Under this assumption the combined 1σ error is equal to the root-sum-square of the various error components. For the specific error components listed at the top of Figure 4.5, this combined error amounts to 18.4 feet (RSS).

The error components for C/A-code navigation are represented by the bar charts at the bottom of Figure 4.5. Notice that the satellite clock and ephemeris errors are identical to the ones for P-code navigation, but the error associated with the ionospheric delay is three times larger because the precise dual-frequency ionospheric correction is not available to the C/A-code users. They must, instead, employ a less accurate polynomial approximation whose coefficients are obtained in real time from the satellite's data stream. The tropospheric error is the same for P-code and C/A-code navigation (6.6 feet), but the receiver noise and resolution and the multipath errors are both 10 times larger because of the lower chipping rate associated with C/A-code navigation.

When selective availability is implemented, the extra ranging error purposely introduced will, on the average, amount to 98.4 feet, and the corresponding total root-sum-square ranging error will equal 116 feet. The C/A-code navigation error, without selective availability, will equal 63 feet.

An approximate value for the 1σ navigation error associated with the Navstar navigation system can be obtained by multiplying the combined total ranging error by the worldwide time-averaged PDOP which, depending on the precise Navstar constellation, amounts to approximately 2.3. When this PDOP value of 2.3 is multiplied by the 18.4-foot P-code ranging error, an average 50-foot navigation error results.

Kalman Filtering Techniques

The various navigation errors quoted so far correspond to a pseudo-ranging "point solution," but, with effective Kalman filtering, substantial error reductions can sometimes be achieved. Kalman filtering is a mathematical technique for combining and smoothing a sequence of navigation solutions to obtain the best real-time estimate of the current position. For stationary users the Kalman filter might involve only four state variables (types of unknowns to be combined and smoothed). The usual selection for the four state variables would be U_x, U_y, and U_z, plus the clock-bias error, C_B.

For an application with moderate vehicle dynamics, an eight-state Kalman filter will usually suffice. Typically, its state variables will consist of the three position coordinates, U_x, U_y, and U_z; the three velocity components, $\dot{U}_x$, $\dot{U}_y$, and $\dot{U}_z$; the clock bias error, C_B; and the clock-bias error rate, $\dot{C}_B$. High-dynamic users typically employ an 11-state Kalman filter with these 8 state variables plus the three acceleration components, $\ddot{U}_x$, $\ddot{U}_y$, and $\ddot{U}_z$.

For high-dynamic applications in which a GPS receiver is integrated with an inertial navigation system, the number of state variables can quickly increase to a much larger number with the addition of accelerometer bias terms, tilt angles, drift-rate terms, and so on.

Military receivers often employ Kalman filtering techniques to compute and display the *figure of merit*, a real-time estimate defining the current error in the navigation solution. A figure of merit of 3, for instance, indicates that the current navigation error lies somewhere between 164 to 246 feet (50 to 75 meters) a figure of merit of 6 indicates a navigation error of 656 to 1,650 feet (200 to 500 meters).

The Navstar receiver determines the current value for its figure of merit by comparing the real-time error statistics coming out of its Kalman filter with previously computed statistics stored in tabular form in the receiver's memory. These previously stored error statistics take into account a large number of parameters and conditions, such as the number of satellites being tracked, the aiding units currently being used, the operating state of the receiver, and so on. With Kalman filtering, the navigation solution typically converges to a substantially improved accuracy level within a few dozen seconds.

In one series of computer simulations, for instance, a military airplane's dynamic maneuvers were simulated with great fidelity as it descended to land on an aircraft carrier steaming forward at a constant speed. During the first 20 seconds of that differential navigation simulation, the relative error between the two moving vehicles converged to within 10 feet. During the next 80 seconds, the relative error shrank to about 3 feet.

Comparable results have been demonstrated under actual flight conditions in which airplanes are flown around racecourse trajectories. In this case the actual location of the plane is determined by mounting corner-cube reflectors on its wings and fuselage to retroreflect ground-base laser beams. Procedures of this type can be used to measure and compare the accuracy of various Kalman filters and the other contributors to the navigation error.

5

User-set Performance

The average positioning error for Navstar navigation is often quoted as a single number, such as 50 feet. But, actually, as Figure 5.1 indicates, at least six grossly different error levels can be specified, depending primarily on which operating mode the receiver uses to obtain its navigation solution. The six modes of operation, which are listed in Figure 5.1, can be partitioned into these two broad categories:

1. Absolute navigation
2. Differential navigation

In the *absolute navigation* mode the receiver determines its position absolutely with respect to a specific set of map coordinates, such as longitude and latitude, as defined in the WGS-84 map coordinate system. In the *differential navigation* mode the receiver determines its position with respect to a fixed base station. The base station picks up the L-band signals from the Navstar satellites in real time and transmits information on its current navigation solution to other, nearby receivers. Within a small local area, differential navigation is considerably more accurate than absolute navigation because many of the errors are common to the two solutions and hence they tend to cancel out.

Accuracy Estimates for Various Methods of Navigation

As the three horizontal stripes in the upper right-hand corner of Figure 5.1 indicate, absolute C/A- and P-code navigation solutions using conventional pseudo-ranging techniques provide three distinct levels of accuracy, de-

THE SIX BASIC LEVELS OF GPS ACCURACY

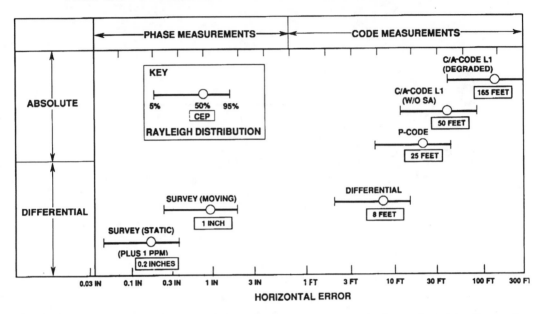

Figure 5.1 The average positioning error of a Navstar receiver depends primarily on the type of signal it is processing (C/A- or P-code), its status with respect to selective availability (degraded or undegraded), and whether or not it is using differential navigation and/or carrier aiding techniques. The six horizontal bars in this figure represent six popular modes of navigation presently being employed by users of various types. Notice that their average (CEP) errors range from 0.2 inches to 165 feet—a variation that spans five orders of magnitude!

pending on the precise method of implementation. Each horizontal stripe spans 90 percent of the navigation errors in a Rayleigh distribution. The Rayleigh distribution function is often used to model radial navigation errors. It resembles the familiar "bell-shaped" distribution curve, with a slight tilt to the right. The small circles near the center of each horizontal stripe mark the most probable navigation error (CEP value) for that particularly mode of navigation. The stripe itself represents ± 45 percent of the cases on either side of the CEP value.

The stripe at the top and to the far right of Figure 5.1 represents the case in which a C/A-code receiver is operated in the degraded mode (with selective availability implemented). In this case the CEP positioning error turns out to be about 165 feet, but under worst case conditions (95 percent probability level), the error can be as large as 330 feet. The undegraded C/A-code solution (without selective availability) is represented by the second bar from the top. In this case the CEP value turns out to be 50 feet, with a "worst case" error of 100 feet. This compares rather favorably with the undegraded P-code solution (third bar from the top), which yields a CEP navigation error of 25 feet.

When a receiver is operated in the differential navigation mode, with relatively gentle vehicle dynamics, a differential navigation solution (fourth bar) typically provides a CEP positioning error of 8 feet or less. In this case, of course, the positioning error is measured with respect to the differential base station not in absolute map coordinates. For slowly moving survey-type solutions with efficient carrier wave aiding (fifth bar), the CEP error shrinks to only about one inch. In a carrier-aided solution, the Navstar receiver uses L_1 and/or L_2 carrier wave measurements to enhance the accuracy of its differential navigation solution. Methods for implementing carrier-aided solutions are discussed in detail in the first half of Chapter 7.

The sixth and final bar in Figure 5.1 represents the case in which carrier-aided navigation solutions are executed by a static user who occupies a fixed survey site for 15 to 45 minutes. In this case the average navigation error can be reduced to only about 0.2 inches, assuming that the receiver and its base station are sufficiently close together. For appreciable separation distances, the surveying error increases by one part per million. This means that the distance in inches between the base station and survey site must be divided by 1 million and then added to the 0.2-inch value to obtain the actual average error. Thus, if base station and survey site are 1 million inches apart (about 16 miles), the total positioning error will amount to 1.2 inches.

Performance Criteria to Consider when Purchasing a Navstar Receiver

When you have decided to purchase a Navstar receiver specifically suited to your particular needs, you should take into account various design parameters and measures of performance before making your final selection. Here are three important measures of performance, each of which is affected by a variety of design decisions:

1. Navigation accuracy
2. Dynamic operating capabilities
3. Jamming immunity

The *navigation accuracy* for a particular receiver provides a quantifiable measure of its likely positioning errors when it is used to obtain a navigation solution. Military specifications for the Navstar navigation system call for a P-code positioning error not to exceed 50 feet CEP. Similar performance specifications for civilian navigation users call for a degraded C/A-code navigation error of 100 meters or less 95 percent of the time. As we have seen in previous paragraphs, differential navigation and carrier-aided solutions can result in much smaller average errors.

The *dynamic operating capabilities* of a Navstar receiver provide us with a convenient numerical measure of its anticipated behavior when it is operated in a vehicle with uncertain dynamics. Single-channel C/A-code receivers often experience degraded operation (or cease to function entirely) when they are subjected to only a few g's. They lose lock on the satellite signals in relatively undemanding dynamic environments, or else they provide inaccurate or erratic navigation solutions. More complicated multichannel P-code receivers can usually operate in high-dynamic environments without serious difficulty.

The *jamming immunity* of a Navstar receiver provides us with a quantifiable measure of its resistance to inadvertent or purposeful (enemy) jamming. P-code receivers usually have much higher jamming immunity than C/A-code receivers. The jamming immunity of both types of receivers can be enhanced by various practical measures, such as terrain masking, null-steering antenna, or careful integration with an inertial navigation system.

Receiver Design Choices

Once you have decided to design or purchase a Navstar receiver, you will need to focus your attention on certain specific design choices in attempting to obtain the most suitable unit at the most affordable price. These design choices, which are discussed in the next few paragraphs, can be grouped into four broad categories:

1. Number of channels and sequencing rate
2. Access to selective availability signals
3. Available performance enhancement techniques
4. Computer processing capabilities

Each of these four categories is further subdivided and explored in the next four subsections. Then, in a later section, their various impacts are summarized in a convenient tabular format.

Number of Channels and Sequencing Rate

As Chapter 4 indicates, three major types of receivers are currently available in the commercial marketplace:

1. Continuous-tracking receivers
2. Slow-sequencing receivers
3. Fast-sequencing receivers

Continuous-tracking receivers are typically rigged with 4 to 12 parallel processing channels, each of which devotes itself to a particular satellite at a particular time. While it is tracking a given satellite, the continuous-tracking receiver measures the signal travel time and the instantaneous Doppler shift. It also decodes the satellite's 50-bit-per-second data stream to gain access its ephemeris constants, its clock correction factors, and any relevant health status information.

A continuous-tracking receiver with four or more channels can provide accurate navigation, high-dynamic operation, and good jamming immunity, but, compared with the other two options, it tends to be costly and complicated. "All-in-view" receivers are even more complicated because they perform their navigation solutions using pseudo-ranging measurements from all of the satellites currently situated above the horizon. An all-in-view receiver typically provides a 20-percent reduction in average navigation error, compared with a receiver that processes the signals from only four satellites. Unfortunately, the largest error reductions occur when several satellites are in view and the solution errors are fairly small. When coverage is skimpy, the all-in-view receiver does not provide much improvement.

Slow-sequencing receivers are considerably cheaper than continuous-tracking receivers, but their performance is degraded in the presence of moderately strong jamming signals, and they do not operate well in high-dynamic environments. For some applications a slightly more complicated two-channel receiver can provide a reasonable design compromise. The second channel can "rove" among the available satellites to establish their signal characteristics and gain access to their current ephemeris constants. Thus, only minimal delays are experienced when a new satellite must be phased in to take over for one that has passed below the horizon or moved into an unfavorable location with bad geometry.

Fast sequencing receivers also use time-sharing techniques, but they dwell on each satellite for an extremely brief interval (typically one two-hundredth of a second) before switching to the next satellite in sequence. "Multiplexing" receivers of this type provide an interesting compromise between the continuous-tracking and the slow-sequencing designs. They can operate under fairly high dynamic conditions, and their somewhat shorter time-to-first-fix is acceptable in many situations. Unfortunately, multiplexing design makes such a receiver considerably more susceptible to enemy jamming, and, hence, it tends to be a poor choice for high-performance military applications.

Access to Selective-availability Signals

When building or purchasing a Navstar receiver, the serious user must determine whether it will be rigged with L_2 signal processing capabilities. If it is available, the L_2 signal can yield dual-frequency ionospheric corrections that improve the accuracy of that particular component of the pseudo-ranging error budget by a factor three compared with the polynomial correction provided by a simpler L_1 receiver. Of course, becoming an authorized user with access to the L_2 signal may be quite difficult. So far, only military users and a few others who can justify their needs have been granted access to the P-code signals—which are the only pseudorandom pulse sequences available on L_2.

The error growth rate for very-long-baseline surveying could be substantially reduced if interferometry-type receivers could be rigged to process the L_2 signals. With a single-frequency L_1 receiver, typical surveying errors amount to about 0.2 inches plus 1 part per million. But, if both the L_1 and the L_2 signals are available, the error growth rates are typically reduced to 0.2 inches plus only 0.5 parts per million. The on-site averaging time for surveying solutions can also be reduced when the L_2 signal is available. A benchmark that requires 45 minutes of averaging time using the L_1 signal typically requires only 15 minutes of averaging time when both the L_1 and the L_2 signals can be used in obtaining the navigation solution.

Code selections must also be taken into account when selecting the number of channels and the sequencing rate for a Navstar receiver. Three fundamentally different approaches are available:

1. C/A-code compatibility
2. P-code compatibility
3. Code-free navigation solutions

Each of these three different selections has its own relative advantages and disadvantages.

C/A-code receivers are cheap and simple. They work well in low-dynamic environments where purposeful or inadvertent jamming signals are not a problem. However, when military encryption is imposed on the L-band signals, the C/A-code selection is the only practical choice for most users because, in that case, the P-code transmissions will be available only to authorized (military) users.

The P-code has a higher chipping rate than the C/A-code, so it provides smaller navigation errors, better jamming immunity, and improved multipath rejection. To obtain the full accuracy of the Navstar system, military P-code users must gain access to the proper encryption keys to eliminate the effect of anti-spoofing and selective availability.

Code-free (carrier-aided) receivers, which are also called "interferometry" receivers, skirt this worrisome thicket of difficulties by making use of their own base stations to gain access to the extra accuracy provided by the L_1 and L_2 carrier waves. Under favorable conditions, codeless interferometry receivers can provide extreme accuracy. Unfortunately, they suffer from severe performance penalties in terms of the signal-to-noise ratio. For this reason an interferometry receiver tends to be a poor choice for a demanding military mission.

Selective availability was turned on for the Block II satellites. Then it was turned off during the Persian Gulf War to allow coalition forces to use plentiful and inexpensive civilian receivers. Then it was turned back on again when the war was over. Consequently, today's authorized users must have access to the proper electronic equipment and the necessary encryption keys in order to restore the full P-code accuracy of the Navstar system. If the proper encryption keys are not available, the positioning accuracy will be degraded—unless U.S. military experts should, for some reason, decide to turn selective availability off again.

Available Performance Enhancement Techniques

A number of techniques are available for enhancing the performance capabilities of a Navstar navigation receiver. These include the use of differential navigation, pseudo-satellites, and aiding inputs from barometric altimeters, atomic clocks, and integrated navigation systems—all of which can substantially enhance the reliability and robustness of the navigation solutions. Of course, these performance enhancement techniques can be implemented only if the receiver has been designed to accept the necessary inputs.

If, for instance, a receiver is to be operated in the differential navigation mode to obtain improved accuracy, it must be equipped with dedicated software and special interface ports for receiving the pseudo-range corrections from the differential base station, most likely in the formats specified by Special Committee 104.

Special Committee 104 was formed by the U.S. Department of Transportation, and its recommended data exchange protocols have been widely reported in *The Journal of The Institute of Navigation* (Washington, D.C.).[1] The Committee includes government and industry participants who develop guidelines to help foster widespread adoption of differential correction techniques. The Committee's most important assignment is to devise standardized data exchange protocols and signal formats for the differential navigation messages. Their message format calls for the broadcast of a 50-bit-per-second data stream from each differential base station, using protocols that are surprisingly similar to the data stream broadcast by the Navstar satellites. Most of the time the differential base station broadcasts real-time pseudo-range corrections, but occasionally it interleaves other types of information.

In order to achieve full differential navigation accuracy, navigators must use continuous-tracking multichannel Navstar receivers. Dual-sequencing receivers are not nearly as accurate when they are used in connection with differential navigation. In most cases the differential navigation solutions recommended by Special Committee 104 are considerably more accurate than the usual absolute navigation solutions only if the user is within two or three hundred miles of the differential base station. The users must receive and process the differential navigation corrections quickly, because their effectiveness degrades with time. These and other issues are further discussed in Chapter 6.

---

[1]The address and phone number of the Institute of Navigation are listed in Appendix C.

Pseudo-satellites are "false" satellites that sit on the ground at fixed locations and transmit navigation signals similar to the ones transmitted by the Navstar satellites. This approach can substantially improve the user's navigation accuracy, especially altitude accuracy. Properly equipped pseudo-satellites can also help insure the integrity of the signals streaming down from the GPS satellites orbiting overhead. Specially designed Navstar receivers with the proper hardware and software modules are required if the user is to take advantage of the signals being broadcast by the ground-based pseudo-satellites. Recommended signal specifications and data formats for pseudo-satellite transmissions have also been developed by Special Committee 104. Pseudo-satellites transmit in the L-band portion of the frequency spectrum, but, unlike the L-band signals from the satellites, they employ pulse-position modulation. In pulse-position modulation the transmitter is activated (transmitting) only a small fraction of the time. This minimizes the jamming of nearby receivers.

Even if a Navstar receiver is not designed to use the pseudo-satellite signals, it should be designed for *pseudo-satellite immunity*. Otherwise, inadvertent jamming of the satellite signals may occur whenever the user approaches one of the pseudo-satellites. The signal formats and the message exchange techniques recommended by Special Committee 104 have been carefully structured to help minimize interference from pseudo-satellites. Nevertheless, any individual who buys a modern Navstar receiver should obtain assurances of pseudo-satellite immunity before making a commitment for final purchase.

Computer Processing Capabilities

A well-designed Navstar receiver must include appropriate software modules to enhance its accuracy, stability, and reliability. Powerful and robust Kalman filtering techniques and satellite selection algorithms are especially important. A properly designed Kalman filter should include static operating capabilities, together with properly programmed adjustments for operation in different acceleration environments. It should also be able to solve for velocity, make proper use of altitude aiding, and provide frequent updates to refresh the Kalman filter before its gains and constants become stale. Kalman filter design is both science and art, and different applications and operating environments call for grossly different Kalman filters.

The satellite selection algorithm should be refreshed at least once per minute. Each time it is refreshed it should take into account the health status of all the available satellites and a large array of GDOP (Geometrical Dilution of Precision) values. All-in-view receivers, which pick up and process the L-band signals from all the visible satellites, should examine a similar array of factors in determining which (if any) navigation signals they choose to ignore and what weights they should assign to the signals they choose to use in their overdetermined solutions.

External navigation aids, such as altimeters, inertial navigation systems, atomic clocks, and other ground-based and space-based navigation systems, can greatly improve the stability, the accuracy, and the jamming immunity of a real-time navigation solution. External navigation aids can also help increase the receiver's integrity and reliability, thus opening up many demanding uses for the Navstar navigation system, including automotive position-fixing and air traffic control that might, otherwise, be too hazardous or too politically sensitive for widespread adoption.

Receiver Design Smart Card

The "smart card" in Table 5.1 summarizes some of the major design choices that should be considered when purchasing or constructing a new Navstar receiver. Each column highlights the specific effects of ten design choices on the accuracy, the dynamic response, and the jamming immunity of the various kinds of Navstar receivers available in today's competitive marketplace.

Notice that all ten of the design choices impact the accuracy of the navigation solution, with the exception of pseudo-satellite immunity, which allows a nonparticipating receiver to operate in the vicinity of pseudo-satellites. By contrast, only four of the design choices impact the jamming immunity and only three of them impact dynamic response: (1) the number of channels, (2) Kalman filtering and update rate, and (3) aiding inputs. One way to benefit from this smart card is to decide which of the basic measures of performance—accuracy, dynamic response, jamming immunity—are important to you, and then go down the appropriate column to pick out those specific design techniques that might improve your particular situation. A more detailed review of the relevant parameters can then help point the way toward various methods for achieving your performance goals at the most affordable price.

Today's Available Navstar Receivers

Forty large companies throughout the world are today producing and marketing Navstar receivers. All together, they offer more than 100 different models, 75 percent of which can provide civilian C/A-code positioning solutions.

Land-based navigation services are provided by about 77 percent of the commercially available receivers. Marine and aeronautical applications are served by 63 and 57 percent of them, respectively. Although the Navstar constellation is financed by military dollars, only 15 percent of today's receivers are designed for military applications.

The most popular units are 1-, 2-, and 5-channel models, but channel numbers ranging from 1 through 12 are all available—with the possible

Table 5.1 Smart Card Design Choices

Design Parameters	Accuracy	Dynamic Response	Jamming Immunity
1. No. Of Channels And Sequencing Rate	Mulit-Channel And Multiplexing Receivers More Accurate In Dynamic Situations	Multi Channel Receivers More Accurate In Dynamic Situations	Multiplexing Receivers More Suceptable To RF Jamming
2. L₂ Capability	L_2 Receivers Provide Dual-Frequency Ionospheric Correction L_2 Receivers More Accurate For Surveying		
3. Code Selections C/A, P, or Codeless	P-Code Receivers More Accurate Codeless Receivers Substantially More Accurate		P-Code Receivers More Immune To Jamming Codeless Receivers More Suceptable To Jamming
4. Access to SA Signals	Access To SA Keys Restores Full P-Code Accuracy		
5. Differential Compatibly	Differential Navigation Provides Substantial Accuracy Improvements		
6. Pseudo-Satellite Compatibility	Pseudo-Satellite Navigation Provides Substantial Accuracy Improvements		
7. Pseudo-Satellite Immunity			Design-Immunity To Pseudo-Satellites Prevents Inadvertent Jamming
8. Kalman Filtering And Update Rate	Sophisticated Kalman Filtering Improves Nav. Accuracy And Flexibility	Sophisticated Kalman Filtering With Frequent Updating Improves Dynamic Response	
9. Satellite Selection	Frequent Updating Of Satellite Selection Improves Nav. Accuracy		
10. Aiding Inputs	Real-Time Aiding With Altimeters, Atomic Clocks, etc, Improves Nav. Accuracy	Velocity Measurements With Inertial Navigation System Can Improve Dynamic Response	Velocity Aiding With Inertial Navigation System Can Improve Jamming Immunity

72

exception of an 11-channel receiver. Some hand-held units weigh only a few ounces each, whereas some larger models weigh as much as 100 pounds or more. The heaviest units serve as base stations for differential navigation, so they are both receivers and transmitters. For this application, miniaturization is not particularly desirable because a differential base station sits permanently at one location on the ground. Power levels of today's receivers range from fractions of a watt up to 200 watts or more. The power-hungry units are, most often, differential navigation transceivers.

Various types of interface ports are available, but the most popular version is the RS-232 interface port. The IEEE-488 Centronics port is also widely used, as is the KYK-13 military interface port, which is equipped to handle encrypted inputs.

Hand-held Receivers

A typical Navstar receiver is about the size of a portable electric typewriter. But, in general, modern versions are getting smaller and smaller, and hand-held units are also being introduced by a number of manufacturers. The first hand-held receiver was affectionately nicknamed the "Virginia Slim" because it was about the same size as a king-size pack of cigarettes. This pioneering device was developed in Cedar Rapids, Iowa, at the Collins Division of Rockwell International under a special contract to DARPA (Defense Advanced Research Project Agency). DARPA researchers frequently sponsor high-risk research projects with the potential for big military payoffs. The Virginia Slim is a 2-channel, dual-frequency P-code receiver weighing about 8 ounces (not counting battery weight). It uses the latest custom-designed gallium-arsenide circuit chips. Chips made from gallium-arsenide are costly and difficult to produce, but they have unusually high switching speeds and good resistance to nuclear radiation.

Another popular hand-held unit, the Magellan Nav 1000 GPS receiver, was originally designed for use aboard pleasure boats, but it was widely used in Saudi Arabia and Kuwait by coalition forces fighting in the Persian Gulf. The Magellan Nav 1000 (see Figure 5.2) weighs 28 ounces and is powered by six AA alkaline batteries. During the Persian Gulf War it retailed for less than $3,000 per unit with a one-year guarantee.

The Nav 1000 features convenient alphanumeric displays with 50 user-supplied waypoints. At the push of a button it provides the user with current range and bearing to the next waypoint. Magellan's receiver is rigged to compute satellite availability anywhere in the world and to evaluate L-band signal strength and satellite geometry. Also, like a bar of Ivory soap, it floats! The designers built it with positive buoyancy so, if a navigator accidently drops it overboard, it can be retrieved.

Hand-held receivers in a variety of configurations are becoming widely available, including one from Japan's Sony Corporation. Excitement rippled through the marketplace when early newspaper reports indicated that the

THE MAGELLAN GPS NAV 1000 HAND-HELD RECEIVER

RECEIVER CHARACTERISTICS

- **WEIGHT = 28 OUNCES**
- **POWER = 6 AA ALKALINE BATTERIES**
- **PRICE ~ $2,000**
- **GUARANTEE = 1 YEAR**

PERFORMANCE FEATURES

- **50 WAYPOINTS**
- **RANGE AND BEARING TO YOUR DESTINATION**
- **TIME TO GO AND ESTIMATED TIME OF ARRIVAL**
- **CROSS TRACK ERROR**

NAVIGATION MADE EASY

- **AUTOMATIC SATELLITE SELECTION**
- **COMPUTES SATELLITE AVAILABILITY ANYWHERE IN THE WORLD**
- **WAKE-UP ALARM TO TAKE POSITION FIX**

Figure 5.2 Magellan's Nav 1000 hand-held receiver, a civilian maritime unit, was widely used by coalition forces during the war with Iraq. It features simple alphanumeric displays, waypoint navigation with 50 user-supplied longitude/latitude waypoints, and a light-weight, compact design. Powered by six ordinary AAA alkaline batteries, the Nav 1000 sold for less than $3,000 at marine supply houses during the Persian Gulf war.

Sony's 4-channel C/A-code receiver would retail for $690, or about one-quarter the cost of competitive hand-held units. Unfortunately, later marketing announcements from Sony fixed the introductory price at $1,300 per unit. Trimble Navigation, Columbia Positioning, and Standard Electrik in Germany are also producing hand-held devices for the personal navigation market.

Commercially Available Navstar Chipsets

Most commercially available Navstar receivers are full-blown devices complete with L-band antennas, software routines, power sources, and control display units. But a few of them are being sold as "chipsets," which include only the solid-state electronic devices that fit inside Navstar receivers. Chipsets are sold primarily to other manufacturers who integrate them with their own antennas, batteries, and other modules before selling them to the end user.

The Navcore V, being marketed by Rockwell International, provides an illustrative example. Navcore V, which is about the same size as an ordinary playing card, features the latest gallium-arsenide technology with custom-integrated circuit chips. It includes 2 megabytes of read-only memory with two low-density random-access circuit chips. When integrated with other appropriate modules, it functions as a 2-channel Navstar receiver. Single Navcore V chipsets sell for $450 each, with mass production quantities available for only $225 per unit.

Magellan Systems Corporation has produced a slightly larger 5-channel GPS receiver module on a two-board set pinned back-to-back. Magellan's device can operate in a 2-g acceleration environment at a velocity of 1000 miles per hour while providing a position update once per second throughout its flight.

Motorola's 6-channel parallel receiver provides one position update per second in a 4-g environment at a velocity of 670 miles per hour. It can be powered by a 1.5-watt DC power supply operating at 5 volts, or it can be plugged into any standard automotive electrical system.

Navstar Electronics in Sarasota, Florida (a subsidilary of Navstar, Ltd., in Daventry, England), markets the Navstar XR4-PC Insertion Card, which can be plugged directly into an IBM Personal Computer (AT or XT model or an equivalent IBM-PC clone). The Navstar Insertion Card draws its power from the PC motherboard, but, when it is operating, it does not disturb the normal operation of the computer. Once it has been installed, the XR4-PC functions as a 2-channel multiplexing receiver capable of tracking up to eight GPS satellites. The position, velocity, and timing measurements it provides can be fed directly into the computer for processing by any commercially available software routines (spreadsheets, databases, etc.) or by custom software developed by the user. A number of imaginative applications will likely emerge once this cleverly designed device is widely available in universities, laboratories, and research centers throughout the technological world.

6

Differential Navigation and Pseudo-satellites

The Navstar GPS is considerably more accurate than most other radio navigation systems. But, even with P-code precision, its pseudo-ranging solutions do not offer sufficient accuracy for some demanding applications, such as carrier landing operations and maritime navigation in inland waterways and constricted harbors. Fortunately, a number of large research groups, including the United States Coast Guard, are attempting to perfect differential navigation to enhance the system's accuracy for these and many other equally demanding applications.

A differential navigation system uses two Navstar receivers that continuously exchange navigation information with one another in real time. One of them acts as a "base station"; the other navigates relative to the base station's location. In effect, the mobile receiver differences between the two real-time navigation solutions, thus eliminating many of their common errors. Satellite ephemeris and clock-bias errors, for instance, tend to be common to the two solutions. To some extent commonality also exists between the ionospheric and tropospheric delays, which are induced as the L-band signals travel from the satellites down toward the ground. The few remaining line-of-sight ranging errors that are not common to the two solutions tend to be relatively small.

Recent tests indicate that, in some cases, differential positioning errors as small as 3 to 6 feet or smaller can be achieved. The actual error depends on the distance between the two receivers, the speed with which the relevant data can be exchanged, and the sophistication of the computer processing techniques, including, of course, the necessary Kalman filters.

Performance Comparisons: Absolute and Differential Navigation

Early computer simulations using an eight-state Kalman filter to model the behavior of a military jet landing on an aircraft carrier confirmed that the differential navigation errors could be as small as 3 to 6 feet. This result was obtained using theoretical computer situations, but actual flight tests with differential navigation have demonstrated that it accurately reflects the real-world situation.

Figure 6.1 compares the accuracy of absolute and differential navigation for a flight in which an airplane was flying around an oval-shaped "racecourse" trajectory. Notice that, during the initial portion of the flight when absolute navigation was being used, the navigation errors were approximately 120 feet in altitude and a little over 60 feet in the horizontal plane. When differential navigation was implemented, both errors were quickly

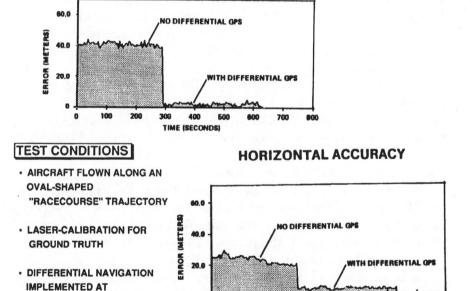

Figure 6.1 In this test series a special airplane was flown around an oval "racecourse" trajectory to test the comparative accuracies of absolute and differential navigation. Both types of navigation solutions were calibrated using precision laser-ranging devices located on the ground. For the first 300 seconds, absolute GPS navigation provided average errors of approximately 120 feet in altitude and 60 feet in the horizontal plane. At t = 300 seconds, differential navigation was implemented. The relative error was then quickly reduced to only about 10 feet.

reduced to 10 feet or less. In recent years a number of similar tests, using various types of hardware and software, have achieved comparable results.

Special Committee 104's Recommended Data-exchange Protocols

In order to encourage widespread adoption of differential navigation, Special Committee 104 has released special data exchange protocols and message formats for the differential navigation corrections.[1] The standardized formats recommended by the committee will allow the differential corrections to be transmitted from base stations at various locations to nearby users in a form they can accept and process as they move from the vicinity of one differential base station to another. The signal formats chosen by the Committee are, in many respects, similar to the navigation data stream broadcast by the Navstar satellites. In particular, the Committee's recommendations call for a 50-bit-per-second data stream using phase-shift-key modulation to mark the boundaries between binary 0s and 1s. The parity checking schemes are also similar, but certain crucial differences do exist between the two types of messages. For one thing, the differential corrections employ a variable word length format. Variable length data words utilize the available bandwidth more efficiently, but the resulting messages are a little more complicated because extra bits must be inserted in the transmission to signal the user when the end of a data word has been reached.

Another important difference is that the message exchange protocols for differential navigation are not limited to a single message type. They include, in fact, a total of 16 different message types. Most of the time the base station transmits pseudo-range corrections (to be explained later). The other 15 message types are occasionally interleaved with these routine differential corrections to provide the user with other useful bits of information, such as the current status of the Navstar constellation, the locations of currently available pseudo-satellites, and the like. The 16 different message types are listed in Figure 6.2, together with a detailed subframe-by-subframe breakdown of message type 1, the primary message type that features pseudo-range corrections and range-rate corrections.

As the sketches in Figure 6.3 indicate, the differential base station obtains its pseudo-range corrections by comparing its conventional real-time pseudo-range measurements with another kind of pseudo-range estimate. The second type of pseudo-range estimate is obtained by subtracting its pre-surveyed location from the satellite's known location. These real-time

---

[1]Special Committee 104 was instituted by the Radio Technical Committee for Maritime Applications (RTCM), which is managed by the Department of Transportation. Its recommendations are strongly influenced by other government bureaus, consultants, and industry representatives.

DATA FORMATS FOR DIFFERENTIAL NAVIGATION

FORMAT SIMILAR TO GPS DATA STREAM

- 50 BITS/SEC
- 16 MESSAGE TYPES
- TYPE-1 MESSAGES INCLUDE:
  - PSEUDO-RANGE CORRECTIONS AND PSEUDO-RANGE RATES
- VARIABLE-LENGTH FORMAT
  (GPS USES FIXED-LENGTH SUBFRAMES)

TYPE NO.	MESSAGE TYPE
1	Differential Corrections
2	Delta Differential Corrections
3	Station Parameters
4	Surveying (Carrier Phase)
5	Constellation Health
6	Null Frame
7	Beacon Almanacs
8	Pseudolite Almanacs
9	High Rate Differential Corrections
10	P-Code Differential Corrections (Reserved)
11	C/A-Code L1, L2 Delta Corrections (Reserved)
12	Health Message (ASCII String)
13-15	Undefined
16	Special Message (ASCII String)

SOURCE:
"SPECIAL COMMITTEE 104 RECOMMENDATIONS FOR DIFFERENTIAL GPS SERVICE."
KALAFUS, VAN DIERENDONCK, PEALER. PROCEEDINGS OF THE 42nd ANNUAL INSTITUTE
OF NAVIGATION MEETING: SEATTLE, WASHINGTON, JUNE 24 - 26, 1986.

TYPE 1 MESSAGE

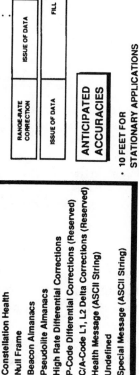

ANTICIPATED ACCURACIES

- 10 FEET FOR STATIONARY APPLICATIONS
- 16 TO 60 FEET FOR DYNAMIC APPLICATIONS

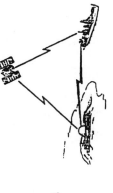

Figure 6.2 Special Committee 104's recommended data exchange protocols for differential navigation include 16 different message types. Most of the time the station broadcasts message type 1, which features pseudo-range corrections and range-rate corrections. The other 15 message types—which provide supplementary information such as base station locations and health status messages—are occasionally interleaved with message type 1.

HOW ARE THE REAL-TIME PSEUDO-RANGE CORRECTIONS (ΔRi' S) DETERMINED BY THE DIFFERENTIAL BASE STATION?

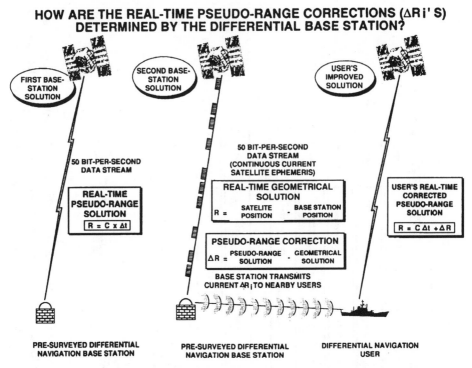

Figure 6.3 A differential base station helps nearby users improve their navigation accuracies by sending them real-time pseudo-range corrections (ΔRi's). The base station determines these corrections by solving for the range to the satellite in two different ways, and then differencing between the two. In the first range solution it measures the signal travel time multiplied by the speed of light. In the second range solution it subtracts the satellite position (computed by using the ephemeris constants) from its pre-surveyed location. Then it differences between the two to get the pseudo-range correction.

pseudo-range corrections are then transmitted to each nearby Navstar receiver, which algebraically adds the appropriate corrections to its current pseudo-ranging solution. This allows it to obtain an improved estimate of its current position.

Field test results indicate that Special Committee 104's data exchange protocols yield differential navigation errors as small as 10 feet for essentially stationary users. For moving users with reasonably gentle dynamics, 15 feet is a more realistic average error. These positioning errors are a little larger than the 3- to 6-foot theoretical errors previously quoted, but they are still substantially smaller than the errors associated with absolute navigation, especially the absolute errors obtained when using degraded C/A-code signals. When differential navigation is employed, it greatly reduces the navigation error. Properly implemented, it removes most of the degradation of error that military experts purposely introduce with selective availability. Of course, differential navigation achieves this favorable result only in the vicinity of a differential base station.

The Coast Guard's Differential Navigation System Tests

In order to evaluate the effectiveness of differential navigation and to determine the influences of ionosphere delays and multipath errors, the U.S. Coast Guard has conducted a series of experiments in which they transmitted, received, and processed differential corrections in the specific formats recommended by Special Committee 104. Coast Guard officials also attempted to determine the effects of various data exchange time delays and separation distances on the accuracy of their differential navigation solutions.

A Trimble 4000A navigation receiver and a Magnavox 5-channel T set were used by the Coast Guard for both static and dynamic testing. Their base station was situated at Boston near the offices of the Department of Transportation. The differential navigation receivers were tested in two distant cities: Groton, Connecticut, and Peoria, Illinois, with baselines spanning 100 and 1000 miles, respectively.

The static navigation test resulted in a mean radial horizontal error amounting to 5.9 feet, with a standard deviation of 0.65 feet. The dynamic test produced a mean radial error of 5.6 feet with a standard deviation of 1.65 feet. Thus, the predicted 7-foot static navigation error was easily achieved.

Dynamic navigation tests were also conducted in which a high-speed patrol boat was piloted through Boston Harbor along a big looping figure-8 trajectory about 3,300 feet long. Ground-truth calibration measurements were obtained from a Falcon 4 Mini Ranger, a short-range ground-based radionavigation system often used in connection with dynamic testing.

Table 6.1 highlights the unmodeled pseudo-ranging errors observed by Coast Guard researchers for the two different baseline lengths. Notice that, with a 100-mile baseline, the unmodeled error in the ionospheric delay ranged from 1 to 2 feet. When the baseline was increased to 1,000 miles, the ionospheric delay created a 4- to 6-foot error. The corresponding values for the unmodeled multipath error were 4 feet and 4 to 6 feet for the 100-mile

Table 6.1 The pseudo-range errors associated with differential navigation for two different baseline lengths

Error source	100-Mile baseline	1,000-Mile baseline
Typical ionospheric pseudo-ranging error	1 to 2 feet	4 to 6 feet
Typical multipath pseudo-ranging error	4 feet	4 to 6 feet
Combined pseudo-range error (RSS)	4.12 to 4.47 feet	5.65 to 8.48 feet

and the 1,000-mile baselines, respectively. These and other, similar test results have helped Coast Guard researchers predict the overall accuracies of various proposed differential navigation systems, determine the required data exchange rates, and establish the maximum practical separation distances between the base stations and users who can benefit from the differential navigation corrections they provide.

Motorola's Mini Ranger Test Results

Clever and effective differential navigation testing of a different sort was recently carried out by researchers at Motorola's production facility in Tempe, Arizona. In that test series company engineers mounted their differential navigation receiver on an electrically powered golf cart and then drove it along great, careless loops in the company parking lot. Calibration for the Mini Ranger's differential navigation system was accomplished with the Falcon 4—a ground-based radionavigation system that has proven to be highly accurate over short operating ranges. Motorola's Mini Ranger can be rigged to operate in the differential mode or, if desired, in the absolute tracking mode. In the test series conducted in the Tempe parking lot, the differential mode was exclusively used.

Before the comparative tests began, the operator drove the electrically powered golf cart in the "locked wheel" mode, which caused it to trace out a tight circle 6 feet in diameter. Later he drove the cart along a zig-zagging trajectory that carried it alternately between the squat concrete car barriers in the company parking lot. One of the test engineers, a Disleyland enthusiast, later dubbed this method of testing "Mr. Toad's Wild Ride" maneuver (see Figure 6.4).

Positioning errors in the differential navigation solution, compared with the Falcon 4 calibration measurements, typically amounted to only about $\pm$ 1.6 feet in each of the three mutually orthogonal coordinate axes. Once they had carefully analyzed the data, Motorola engineers were convinced that, by carefully focusing their attention on the systematic oscillations of the vertical component, they could detect the small undulations created by the drainage ditches in the company parking lot.

The Mini Ranger navigation system provides a position update once per second, with a time-to-first-fix of approximately 2 minutes. At the user's option it can operate in the absolute GPS navigation mode in 13 different coordinate systems, including WSG-84, the 1916 Fisher Ellipsoid, and the Australian National Datum Plane.

Similar tests were conducted with a GPS receiver mounted on a gasoline-powered Minivan cruising through the streets of Noordwijk in the Netherlands. When the receiver was operated in the absolute navigation mode, its position fixes were displaced by several dozen feet from the highway, thus conclusively demonstrating that the absolute navigation solutions were exhibiting substantial errors. But, when that same GPS receiver was operated in the differential mode, its measured ground track stayed near the

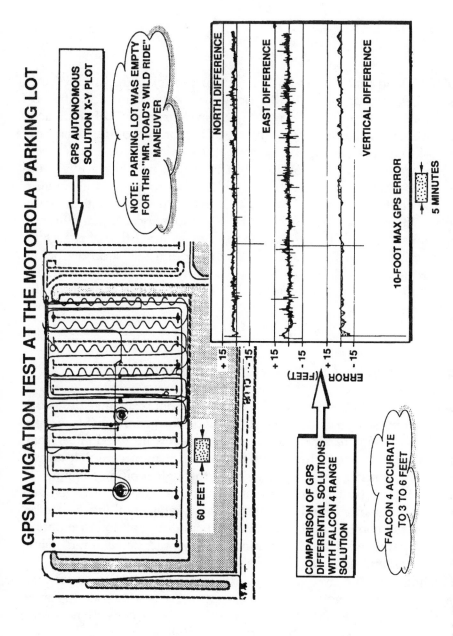

Figure 6.4 "Mr. Toad's Wild Ride" maneuver, as executed in the Motorola parking lot in Tempe, Arizona, demonstrated that differential navigation can produce positioning errors smaller than 10 feet over relatively short baselines. In this test series a differential navigation receiver was mounted on an electrically powered golf cart. The cart was then driven along a zig-zagging trajectory that carried it between alternate pairs of squat concrete barriers in the company parking lot. Accurate ground-truth measurements for the test were provided by the short-range Falcon 4 radionavigation system.

83

center of the highway with obvious precision. Incidentally, during the Noordwijk test series, the Minivan was driven along the highway at 80 miles an hour, a speed that would have been clearly illegal in any counterpart city in the United States.

COMSAT's Data Distribution Service for the Gulf of Mexico

The COMSAT Corporation, a publicly held entity specializing in communication satellites, recently instituted a space-based service in which differential corrections are broadcast to Navstar users in or near the Gulf of Mexico. So far, the users consist primarily of offshore oil exploration outfits operating seismic exploration vessels on the waters of the Gulf.

The differential corrections originate in Houston, but they are relayed through INMARSAT communication satellites hovering over the equator south of the continental United States. COMSAT's tests indicate that any user within 600 miles of Houston can achieve reasonably accurate differential corrections. Initially, the corrections were distributed free of charge to introduce new users to the system, but subscription fees are now being imposed. If the differential correction service turns out to be a profitable enterprise, COMSAT officials intend to duplicate it in other areas throughout the Western world.

Wide-area Differential Navigation Services

The Coast Guard's field test results indicate that the data exchange protocols recommended by Special Committee 104 provide corrections that lose their effectiveness for users that are located more than a few hundred miles from the differential base station. However, by transmitting more complete arrays of information at faster data exchange rates through orbiting satellites, it may be possible to provide differential corrections that are useable over continent-wide distances. Studies indicate that coverage of this type can be provided for the continental United States, using only about 15 monitor stations, compared with the 500 stations that would be required using conventional differential navigation techniques.

The architecture for the wide area differential navigation system is sketched in Figure 6.5. Notice that, in this approach, simple pseudo-range corrections are not employed. Instead, three other types of correction parameters are broadcast in real-time by the orbiting satellite:

1. Satellite ephemeris corrections
2. Clock bias errors
3. Ionospheric correction terms

WIDE AREA DIFFERENTIAL GPS

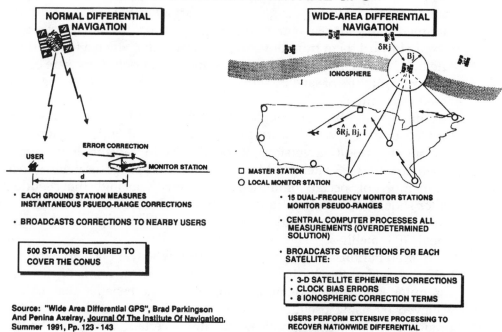

NORMAL DIFFERENTIAL NAVIGATION

ERROR CORRECTION

USER

MONITOR STATION

d

WIDE-AREA DIFFERENTIAL NAVIGATION

IONOSPHERE

δRj

Bj

$\delta \hat{R}j, \hat{B}j, \hat{I}$

☐ MASTER STATION
○ LOCAL MONITOR STATION

* EACH GROUND STATION MEASURES INSTANTANEOUS PSUEDO-RANGE CORRECTIONS

* BROADCASTS CORRECTIONS TO NEARBY USERS

500 STATIONS REQUIRED TO COVER THE CONUS

* 15 DUAL-FREQUENCY MONITOR STATIONS MONITOR PSEUDO-RANGES

* CENTRAL COMPUTER PROCESSES ALL MEASUREMENTS (OVERDETERMINED SOLUTION)

* BROADCASTS CORRECTIONS FOR EACH SATELLITE:

• 3-D SATELLITE EPHEMERIS CORRECTIONS • CLOCK BIAS ERRORS • 8 IONOSPHERIC CORRECTION TERMS

Source: "Wide Area Differential GPS", Brad Parkingson And Penina Axelray, <u>Journal Of The Institute Of Navigation,</u> Summer 1991, Pp. 123 - 143

USERS PERFORM EXTENSIVE PROCESSING TO RECOVER NATIONWIDE DIFFERENTIAL CORRECTIONS

Figure 6.5 Wide-area differential navigation can provide continent-wide differential corrections comparable in accuracy to the corrections provided by conventional differential navigation. However, with this system only 15 widely separated monitor stations are required to cover the continental United States. Three different types of correction parameters must be broadcast by the data distribution satellite: (1) satellite ephemeris corrections, (2) clock bias errors, and (3) ionospheric parameters. When these corrections are used to adjust a real-time navigation solution, the average positioning error shrinks to one-twentieth the comparable error achieved with absolute GPS navigation.

The 15 monitor stations, which are installed at widely separated locations, measure these parameters by picking up the real-time navigation signals from the GPS satellites. Any deviations between the telemetered and the observed satellite ephemeris constants are incorporated in the wide-area differential correction message, together with improved clock corrections and a series of eight constants that define the current continent-wide behavior of the earth's ionosphere.

By applying these three types of corrections to their current navigation solutions, differential navigation receivers all over the country can greatly improve their positioning solutions. Studies indicate that, within a coast-to-coast coverage area, 95-percent error reductions can be anticipated compared with stand-alone GPS navigation solutions.

Pseudo-satellites

Even before any of the Navstar satellites had been launched into space, hundreds of navigation tests had been conducted using GPS-like transmitters positioned on the floor of the desert at Yuma, Arizona. This setup was called "the inverted test range" because airplanes flew over it with belly-mounted navigation antennas rigged to pick up the navigation signals coming toward them from the ground below. Later, when a few of the GPS satellites had been launched into space, navigation solutions were obtained by using a mixture of satellite signals merged with signals from the ground-based transmitters.

Today, ground-based transmitters that mimic the L-band signal format of the GPS are called *pseudo-satellites* (false satellites).[2] Pseudo-satellites broadcast slightly modified versions of the C/A-code transmissions streaming down from the GPS satellites. Many proponents of pseudo-satellites, including members of Special Committee 104, argue convincingly that pseudo-satellites should, in addition, transmit differential corrections to nearby users and also act as monitor stations to help insure the integrity of the conventional GPS satellites.

·Dr. Brad Parkinson at Stanford University, who was once in charge of the GPS Program at the Air Force Space Division, set up some of the earliest computer simulations demonstrating the solid benefits that could be achieved by using pseudo-satellites. Dr. Parkinson became interested in pseudo-satellites when he noticed that, under certain conditions, the 18-satellite GPS constellation would occasionally result in a Vertical Dilution of Precision (VDOP) of 12 or even larger in the vicinity of the San Francisco International Airport. A Vertical Dilution of Precision in that magnitude range would create extremely large vertical (altitude) navigation errors. The resulting uncertainty could be especially bothersome if the Navstar signals were to be used for landing operations at airports, large or small.

Brad Parkinson attempted to circumvent this difficulty by running computer simulations in which he introduced a single ground-based pseudo-satellite in the vicinity of the San Francisco airport. Soon he found that the vertical Dilution of Precision was only about 0.7 with one pseudo-satellite. Thus, the use of a pseudo-satellite would greatly increase the safety and reliability of aircraft landings.

Parkinson relied on his computer simulations to determine the best location for the ground-based pseudo-satellite. In his initial studies, he discovered that the most effective location was 30 miles south of the runway. Why 30 miles south of the runway?

Placing a pseudo-satellite immediately adjacent to an airport runway is a bad idea for two reasons:

1. The pseudo-satellite tends to jam the signals coming down from the GPS satellites circling overhead.
2. Navigation accuracy is degraded compared with positioning the pseudo-satellite 30 miles south of the runway.

---

[2]Some experts shorten the name "pseudo-satellite" to "pseudolite," a cryptic moniker that does nothing but obscure the meaning and confuse the noninitiated.

Parkinson reached this conclusion by trial and error. But later he reasoned that a 30-mile southward location probably produces smaller average errors because the GPS satellites are in 55-degree orbit planes. This places more satellites at high latitudes, so more of them typically lie north of the runway rather than to the south. Thus, a southward location for the ground-based pseudo-satellite tends to provide improved viewing geometry.[3]

Later, during a more detailed study, in which he took a larger number of variables into account—such as radio frequency interference and the local topography—Brad Parkinson found a way to determine the precise optimal location for the pseudo-satellites servicing each specific airport. In this case the best location for a pseudo-satellite is sometimes, but not always, 30 miles south of the runway.

Brad Parkinson is convinced that pseudo-satellite installations can be constructed for $100,000 to $200,000 each, and that America will probably need one or more for each airport of appreciable size. Approximately 6,000 airstrips of various types are in operation in the United States. Most do not have useable landing aids. Pseudo-satellite installations could prove especially beneficial for some of these poorly equipped landing strips.

If the signal structure for the pseudo-satellites is similar to the one employed by the GPS satellites, it can act as an unintended jammer, thus interfering with the proper reception of the L-band signals coming from the GPS satellites. To minimize these jamming effects, some form of time-division multiplexing is necessary, together with careful design for all the GPS receivers to make them more immune to pseudo-satellite jamming.

Special Committee 104's Data Exchange Protocols for Pseudo-satellites

After an extended series of meetings, the members of Special Committee 104 have selected a 1,023-chip Gold Code for the ground-based pseudo-satellites. With the self-jamming problem firmly in mind, committee members have recommended specific pulse-position modulation techniques in which each pseudo-satellite will be active (broadcasting) only about 10 percent of the time. The pseudo-satellite uses the precise timing signals from the GPS constellation to gate itself "on" during scattered 90.91-microsecond intervals (see Figure 6.6). During each gating interval it transmits 93 C/A-code chips so that, after 11 active intervals, its entire 1,023-bit C/A-code has been transmitted. The active intervals are purposely staggered with respect to one another in accordance with the schedule listed schematically on the left-hand side of Figure 6.6.

This specific time-division multiple access technique allows the pseudo-

---

[3]In the southern hemisphere, the pseudo-satellite should, of course, be positioned *north* of the airport runway.

PSEUDO-SATELLITE SIGNAL SPECIFICATIONS

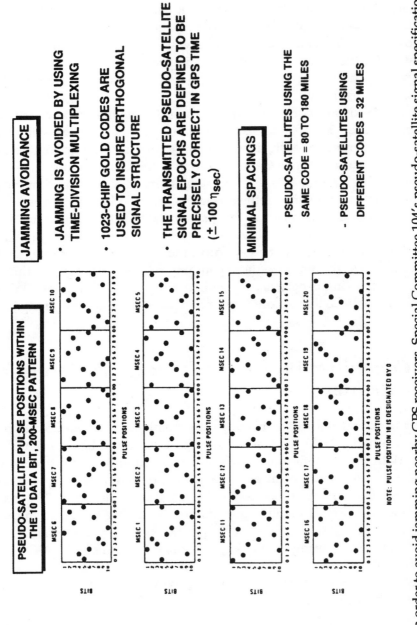

PSEUDO-SATELLITE PULSE POSITIONS WITHIN THE 10 DATA BIT, 200-MSEC PATTERN

JAMMING AVOIDANCE

- JAMMING IS AVOIDED BY USING TIME-DIVISION MULTIPLEXING
- 1023-CHIP GOLD CODES ARE USED TO INSURE ORTHOGONAL SIGNAL STRUCTURE
- THE TRANSMITTED PSEUDO-SATELLITE SIGNAL EPOCHS ARE DEFINED TO BE PRECISELY CORRECT IN GPS TIME ($\pm$ 100 ηsec)

MINIMAL SPACINGS

- PSEUDO-SATELLITES USING THE SAME CODE = 80 TO 180 MILES
- PSEUDO-SATELLITES USING DIFFERENT CODES = 32 MILES

NOTE: PULSE POSITION 10 IS DESIGNATED BY 0

Figure 6.6 In order to avoid jamming nearby GPS receivers, Special Committee 104's pseudo-satellite signal specifications call for the use of pulse-position modulation. With the recommended scheme, each pseudo-satellite transmitter is active only about 10 percent of the time. The pseudo-satellite breaks its 1,023-bit Gold code into eleven 93-chip segments, which are then transmitted during eleven scattered 90.91-microsecond intervals. To avoid mutual interference, pseudo-satellites using different Gold Codes cannot be closer than 32 miles. Those using different Gold Codes must be at least 80 to 180 miles apart.

satellites to transmit their modulated signals only a small fraction of the time so that they create a much smaller jamming hazard for nearby users.

Studies sponsored by Special Committee 104 have shown that two different pseudo-satellites using the same Gold Code cannot be any closer than 80 to 180 miles; otherwise, they will interfere with one another. Two pseudo-satellites using the same Gold Codes cannot be any closer than 32 miles.

The experts from Special Committee 104 have issued practical guidelines for designing and building pseudo-satellite–compatible receivers and pseudo-satellite–immune receivers. A pseudo-satellite–compatible receiver can pick up and use the pseudo-satellite signals to obtain improved navigation accuracy. A pseudo-satellite–immune receiver cannot pick up and use the pseudo-satellite signals, but it can pick up the signals from the GPS satellites, even when it is in the vicinity of an active pseudo-satellite.

The Committee's guidelines emphasize that a successful pseudo-satellite receiver must be able to receive and process the pseudo-satellite signals continuously, so it must be a continuous-tracking multichannel receiver. Of course, the pseudo-satellite receiver must be designed to generate the pseudo-satellite's C/A-code using pulse-position modulation techniques and it must be rigged to perform all necessary computations.

The popular 7-bit ASCII code used in personal computers will be used to encrypt up to four consecutive alphanumeric characters to allow the receiver to figure out which pseudo-satellite is transmitting the Gold Code pulses being received in the L-band portion of the frequency spectrum. The pseudo-satellites near Los Angeles International Airport, for instance, would be denoted by the ASCII symbols LAX1, LAX2, LAX3, and so forth.

Comparisons Between Differential Navigation and Pseudo-satellites

One important advantage of a pseudo-satellite over a differential navigation transmitter is that the pseudo-satellite improves the coverage characteristics and the geometry of the conventional GPS constellation. Pseudo-satellites also provide excellent vertical navigation accuracy. When pseudo-satellites are used, no extra receivers are required. This is true because the (modified) GPS receiver doubles as the pseudo-satellite receiver. In most proposed pseudo-satellite systems, the pseudo-satellite base stations would also transmit real-time differential navigation corrections interleaved with their normal C/A-code transmissions. In addition, they would transmit integrity-related warnings to local users if any of the satellites in the GPS constellation appear to be transmitting inaccurate navigation signals.

One practical disadvantage of pseudo-satellites is that they are limited to line-of-sight coverage. This is true because the transmission must be in the L-band portion of the frequency spectrum which does not reflect off the ionosphere. Pseudo-satellite receivers also require extra hardware modules and software routines to allow them to choose and identify the pseudo-satellites in their vicinity. Finally, if pseudo-satellite transmitters are widely installed, all critical GPS receivers must be pseudo-satellite compatible. Otherwise, the signals from the GPS satellites may be jammed whenever the receiver is close to a pseudo-satellite.

Additional comparisons between differential navigation and pseudo-satellites are provided in Table 6.2. Notice that the choice of transmission frequencies is more flexible for differential navigation because the transmissions are not constrained to be in the L-band portion of the frequency spectrum. The coverage area for differential navigation typically spans a circle 300 to 500 miles in radius centered around the base station. For pseudo-satellites, only line-of-sight L-band coverage is provided. Depending on the height of the transmitting tower, this range is usually 30 miles or less.

The navigation errors associated with a differential navigation solution depend to some extent on the data exchange rate. Errors typically range from 3 to 30 feet. Pseudo-satellites provide roughly comparable navigation errors in the horizontal plane, but their vertical (altitude) errors are considerably smaller. No worrisome user-set jamming problems arise with the implementation of differential navigation, but with pseudo-satellites the transmitters must be properly spaced and the receivers must be carefully designed to minimize the possibility of satellite jamming by the stronger pseudo-satellite signals. This is true even though the pseudo-satellite base stations will be rigged to transmit their signals using pulse-position modulation techniques.

Table 6.2 Comparison between differential navigation and pseudo-satellites

Comparison criteria	Differential navigation	Pseudo-satellites
Transmission Frequencies	Various possibilities	L-band transmissions required
Coverage area	Typically 300 to 500 miles	Line-of-sight coverage only (typically 30 miles)
Navigation accuracy	Depends on data exchange rate (typically 3 to 30 feet)	Similar to differential navigation, superior in the vertical dimension
Integrity monitoring	Usually provided	Usually provided
User-set jamming	No additional difficulties	Jamming can be minimized by careful user-set design

7

Interferometry Techniques

Most of today's receivers use the pseudorandom C/A-and P-code pulse sequences broadcast by the GPS satellites to obtain their current positioning solutions. But a more sophisticated technique called *interferometry* derives information for its navigation solutions from the sinusoidal carrier waves coming down from the satellites. Interferometry solutions, which are also called carrier-aided solutions, are more difficult to obtain, but, in situations where they are valid, they can provide surprisingly large reductions in the navigation errors, especially for static and low-dynamic surveying applications. Some spaceborne applications can also benefit from carrier-aided processing techniques.

The Classical Michaelson-Morley Interferometry Experiment

Interferometry methods first received widespread attention when they were used in the famous Michaelson-Morley experiment, which proved conclusively that the ether did not exist. The ether was a fanciful substance that was believed to carry electromagnetic waves through the vacuum of space. Nineteeth century scientists endowed the ether with a number of semi-magical properties, such as complete weightlessness, total transparency, and infinite rigidity. If the ether existed, it surely carried beams of light along with it in some preferred direction. The earth travels around the sun at 67,000 miles per hour, and the sun whirls around the center of the Milky Way Galaxy at an even faster rate. Only by the most improbable coincidence would an earth-based observer be stationary with respect to the ether.

Michaelson and Morley devised a clever device for measuring the velocity of light in various directions to see how the movement of the ether might affect its propagation speed. Their mechanism broke a beam of light into two parts, sent those two parts along mutually perpendicular paths, and then brought them back together again to check their propagation velocities relative to one another.

First the light was sent through an optical filter and a focusing lens to create parallel rays of monochromatic light (see Figure 7.1). Then it was directed toward a partially silvered mirror that reflected half the light, but allowed the other half to pass on through. The portion that passed through the partially silvered mirror hit a fixed, fully silvered mirror and was reflected back to the surface of the partially silvered mirror. The portion that was reflected by the partially silvered mirror traveled to a movable fully

THE MICHELSON-MORLEY INTERFEROMETER

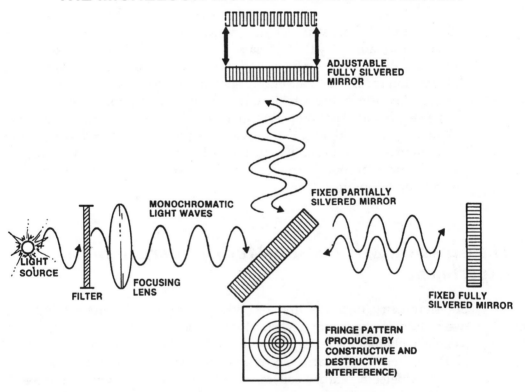

Figure 7.1 Michaelson-Morley's interferometry apparatus uses a half-silvered mirror to divide a beam of monochromatic light into two parts: one part is sent to a fixed mirror, the other is reflected to an adjustable (movable) mirror. The beams then retrace their paths and recombine to form interference fringe patterns—concentric bands of dark and light. When the adjustable mirror is moved up or down one-quarter of a wave length, the dark concentric bands become light and vice versa.

silvered mirror whose position could be manually adjusted by turning two small thumb screws.

Constructive and destructive interference between the two reunited beams created concentric circles of light and dark. Each time the thumb screws were adjusted enough to shorten the path length by one-quarter of a wavelength of the monochromatic light, the dark rings turned to light, and vice versa.

In 1907 Albert Abraham Michaelson was awarded the Nobel Prize for his pioneering work in interferometry techniques. And yet for decades thereafter, the methods that he and his talented colleagues perfected were used for only a few rather esoteric applications. Today, by contrast, interferometry techniques are improving our lives in a hundred dozen different ways, most of which are totally hidden from public view.

Measuring Attitude Angles with Special Navstar Receivers

A specially designed Navstar receiver can make use of simple interferometry techniques to determine its angular orientation with respect to the electromagnetic waves coming down from the GPS satellites. This is accomplished by processing a series of carrier wave measurements from a single satellite picked up by two different user-set antennas separated by a rigid bar. As Figure 7.2 indicates, the carrier waves from a distant satellite travel along essentially parallel trajectories to reach the two antennas. If the rigid bar is tipped at an angle with respect to the wave front, the path lengths followed by the two parallel carrier waves will be unequal. Consequently, if we display both carrier waves on an oscilloscope, they will be displaced with respect to one another. Their phase mismatch can be used to determine the relative orientation angle, θ, which is sketched in the lower left-hand corner of Figure 7.2. Multiple measurements of this type using the L-band signals from various GPS satellites—together with the information they broadcast defining their Keplerian orbital elements—allow the receiver to determine its three independent attitude angles in real time.

A larger separation distance between the two antennas (longer rigid rod) can theoretically increase the accuracy with which the attitude angles can be ascertained. However, larger separation distances also create additional solution ambiguities. Ambiguities in the solution arise from the fact that the receiver cannot distinguish between a pair of path lengths that differ by one-half a wavelength, one and a half wavelength, two and a half wavelengths and so on. Consequently, the angle θ could have a large number of different values. Several promising solutions to this problem are being systematically explored.

PRECISION ANGULAR ORIENTATION MEASUREMENTS USING THE GPS SIGNALS

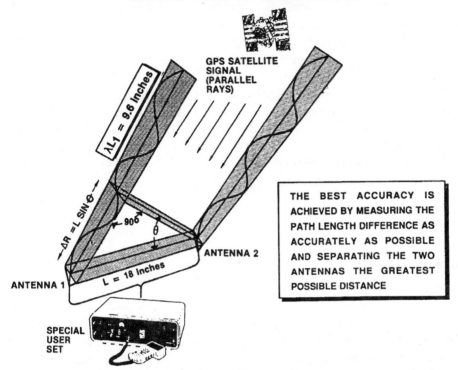

GPS SATELLITE
SIGNAL
(PARALLEL
RAYS)

$\Delta L_1 = 9.6$ Inches

$\Delta R = L \sin \theta$

90°

θ

L = 18 Inches

ANTENNA 1

ANTENNA 2

THE BEST ACCURACY IS
ACHIEVED BY MEASURING THE
PATH LENGTH DIFFERENCE AS
ACCURATELY AS POSSIBLE
AND SEPARATING THE TWO
ANTENNAS THE GREATEST
POSSIBLE DISTANCE

SPECIAL
USER
SET

Figure 7.2   The angular orientation of a rigid bar separating two antennas can be measured by a special GPS receiver that uses interferometry techniques to determine the desired solution. This is possible because the carrier wave must follow a longer path to reach the antenna of the left than it follows to reach the one on the right. Increasing the separation distance between the two antennas improves the accuracy of the device, but larger separation distances also give rise to additional solution ambiguities.

Eliminating Solution Ambiguities

Each Navstar satellite transmits L_1 and L_2 carrier waves that are 9.6 and 7.5 inches long, respectively, so an antenna separation distance of only a few feet can create an enormous number of solution ambiguities. These ambiguities can be resolved, to some extent, by making careful measurements and then using careful computer processing techniques. An alternate approach makes use of an electronically shifted antenna that gradually increases the separation distance between the two antenna phase centers. At first, the two interface ports on the receiver are both fed from the same antenna. Then, gradually, the other antenna feed is electronically shifted along a straight line from one end of a rigid bar to the other. During this interval, the receiver keeps track of the number of wavelengths that have swept by, thus greatly reducing the possibility of unresolvable solution ambiguities. Other promising approaches include software resolution and antennas mounted on the two ends of a rigid rotating rod.

Practical Test Results

An interferometry-type receiver can typically resolve the L_1 and/or L_2 carrier waves to within one-twentieth of a wavelength (about one-half inch). The corresponding theoretical limit on the accuracy of an attitude-measuring receiver rigged with a 5-foot rigid bar is about 10 arc seconds. In actual practice an interferometry-type receiver assembled and tested by Texas Instruments has achieved an accuracy of 30 arc seconds under realistic field test conditions.

Additional attitude-determination experiments have recently been conducted in the Netherlands using a special Navstar receiver. In this case the two antennas were separated by a 6.6-foot rigid rod mounted on a rotating framework. This allowed the experimenters to calibrate the two precisely constructed antennas to reduce any inaccuracies resulting from phase-center offsets in their design, production, and installation.

Bias errors in the software solutions were eliminated by using double-differencing techniques, and any carrier wave ambiguities were reduced by triple differencing of the results. Double and triple differencing techniques are discussed later in this chapter in connection with geodetic surveying applications—which also use interferometry measurements to achieve improved results.

Using Interferometry to Fix Position

One of the earliest companies to market positioning receivers that relied on interferometry techniques for precise surveying was the ISTAC Corporation in Pasadena, California. Dr. Peter E. McDoran, who founded ISTAC, was at that time working on interferometry techniques for Pasadena's Jet Propulsion Laboratory. He was assigned to develop some rather esoteric position-fixing procedures for earthquake fault monitoring and in connection with the orbital positioning for the TOPEX oceanographic satellite.

When Dr. McDoran began to understand the enormous power and versatility of interferometry solutions, he secured permission from NASA to market the new technology to commercial users. The methods he and his colleagues perfected utilize the carrier waves from distant satellites to fix the user's position to a high degree of precision. ISTAC's receivers retail for about $50,000 each, including user instruction and the expensive software routines that yield precise interferometry solutions.

Offshore oil exploration outfits are a major market for interferometry-type GPS receivers. The navigation solutions they obtain are intrinsically differential, so a base station capable of picking up the Navstar signals must be located on a nearby shore. The base station broadcasts information about its current navigation solution to a second (and sometimes a third) receiver located on an offshore oil exploration vessel.

The positioning accuracy of an interferometry receiver increases systematically with the averaging time. This is true because, as the satellites travel across the sky, a larger number of phase angle measurements can be made with different aspect angles. Dr. McDoran estimates that, with a 5-second averaging time, the statistical noise remaining amounts to approximately 3 feet. With a 2-minute averaging time, that error can be further reduced to 9.5 inches. With 20 minutes of averaging, an estimated error of one inch or so will still remain in a typical ISTAC navigation solution.

For some applications involving off-shore oil exploration vessels, the base station transmits information to two different interferometry receivers. One of them is attached to the seismic survey vessel; the other is positioned on the aft end of the trailing hydrophone array, which sticks out from the back of the boat, sometimes to a distance of three miles or more.

Single, Double, and Triple Differencing Techniques

Figure 7.3 highlights various types of differencing techniques used in conjunction with carried-aided (interferometry) navigation solutions. As the "stairstep" sketch at the top of the figure indicates, differencing can be accomplished in three basic ways:

1. Across receivers
2. Across satellites
3. Across time

In accordance with standard conventions adopted by the surveying community, this is the standard computational order—across receivers, across satellites, across time—that is used in connection with single, double, and triple differencing techniques.

The *single difference* (across receivers) can be defined as the instantaneous difference in phase between the received signals as measured by two different receivers simultaneously observing the same satellite. This single differencing eliminates the clock-offset error in the satellite clock because the clock error is common to the two measurements.

The *double difference* (across receivers and across satellites) is obtained by differencing the single differences for one satellite with respect to the single differences for the second (reference) satellite. This double differencing method eliminates both satellite clock offsets and receiver clock offsets because both of these errors are common to the two single differences being subtracted from one another. Both types of clock errors, therefore, cancel in the double differencing process.

The *triple difference* (across receivers, across satellites, and across time) is

SINGLE, DOUBLE, AND TRIPLE DIFFERENCING TECHNIQUES

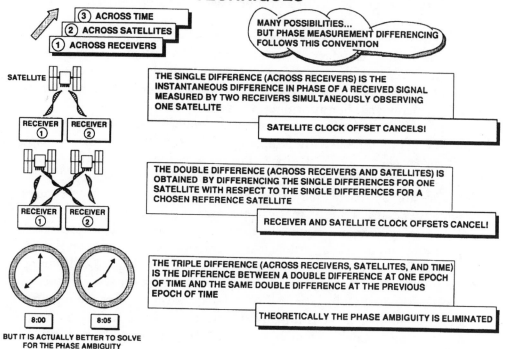

(3) ACROSS TIME
(2) ACROSS SATELLITES
(1) ACROSS RECEIVERS

MANY POSSIBILITIES... BUT PHASE MEASUREMENT DIFFERENCING FOLLOWS THIS CONVENTION

SATELLITE

RECEIVER (1) RECEIVER (2)

THE SINGLE DIFFERENCE (ACROSS RECEIVERS) IS THE INSTANTANEOUS DIFFERENCE IN PHASE OF A RECEIVED SIGNAL MEASURED BY TWO RECEIVERS SIMULTANEOUSLY OBSERVING ONE SATELLITE

SATELLITE CLOCK OFFSET CANCELS!

RECEIVER (1) RECEIVER (2)

THE DOUBLE DIFFERENCE (ACROSS RECEIVERS AND SATELLITES) IS OBTAINED BY DIFFERENCING THE SINGLE DIFFERENCES FOR ONE SATELLITE WITH RESPECT TO THE SINGLE DIFFERENCES FOR A CHOSEN REFERENCE SATELLITE

RECEIVER AND SATELLITE CLOCK OFFSETS CANCEL!

8:00   8:05

THE TRIPLE DIFFERENCE (ACROSS RECEIVERS, SATELLITES, AND TIME) IS THE DIFFERENCE BETWEEN A DOUBLE DIFFERENCE AT ONE EPOCH OF TIME AND THE SAME DOUBLE DIFFERENCE AT THE PREVIOUS EPOCH OF TIME

THEORETICALLY THE PHASE AMBIGUITY IS ELIMINATED

BUT IT IS ACTUALLY BETTER TO SOLVE FOR THE PHASE AMBIGUITY

Figure 7.3 Surveyors employ systematic differencing techniques to eliminate some of the solution ambiguities in their over determined carrier-aided positioning solutions. By single differencing across receivers, they can eliminate any clock offsets in the satellite clock. By double differencing across satellites they can eliminate both receiver clock offsets and satellite clock offsets. By triple differencing across time they can, in theory, eliminate all phase ambiguities. In actual practice, however, polynomial curve-fitting techniques that are applied to a whole series of double differences can often produce superior results.

defined as a difference between the double differences (across receivers and across satellites) at two successive time points. Triple differencing of this type should theoretically eliminate all timing errors and all phase ambiguities. In practice, however, the instantaneous phase offsets are often determined by polynomial curve-fitting of successive double differences. This achieves the same objective in a more stable and reliable way.

The POPS Post-Processing Software

A number of highly sophisticated commercial software routines are available for handling the single, double, and triple differencing, the curve-fitting procedures, and the various other computer processing operations necessary for manipulating large arrays of carrier-aided positioning measurements. In studies of this type, post-processing techniques are often used to further enhance the accuracy of the navigation solutions. In post-processing the Keplerian orbital elements defining the orbits of the GPS satellites are obtained after the fact from a ground-based tracking network for use in a non–real-time navigation solution.

Satellite tracking data obtained in this way is substantially more accurate than the extrapolated ephemeris data that is picked up from the GPS satellites in real time. Post-processing data a few days old can be obtained from the master control station, but several other private data sources are also furnishing similar information to serious Navstar users.

Appendix E shows how you can order unbiased information on surveying receivers complete with the appropriate software routines. One popular software package, the *POst-Processing Software*, or POPS, has been used for several years in connection with precise surveying applications. The POPS routines are capable of analyzing carrier-aided measurements from surveying networks involving as many as ten base stations. An ordinary desk top IBM-PC AT personal computer or a similar PC clone handles the processing operations and displays the data in convenient tabular and pictorial formats.

The POPS processing routines employ single and double differencing techniques to fix the locations of the various base stations, sometimes to an accuracy of one-half inch or less. Any detectable cycle slips are repaired by using polynomial curve-fitting routines at certain key points in the processing operations. Working together in a joint project, researchers at America's Magnavox Corporation and Wild-Heerbrugg in Switzerland programmed and integrated the necessary software modules.

The input-output procedures used in the POPS program are surprisingly user friendly. Among other things, they employ menu listings and special function keys to aid in efficient operation. A special "HELP key" is also provided for any users who may encounter special difficulties.

Spaceborne Interferometry Receivers

A few spaceborne receivers are being designed to fix their positions using carrier-aided interferometry measurement techniques. The TOPEX oceanographic satellite, for instance, will be rigged to determine its orbital elements to within 4 inches using real-time carrier-aided measurements of the L-band carrier waves broadcast by the GPS satellites.

TOPEX, which is being sponsored jointly by the Jet Propulsion Laboratory and the French government, is designed to measure the surface contours of the oceans to a high degree of accuracy: 3/8 to 3/4 of an inch! This will be accomplished by reflecting radar pulses off the water below, and then measuring the two-way signal travel time (see Figure 7.4).

What is the purpose of mapping ocean contours with such fantastic precision? One important aim is to learn more about the El Nino phenomenon—a meteorological aberration that brings freakish weather to vast areas

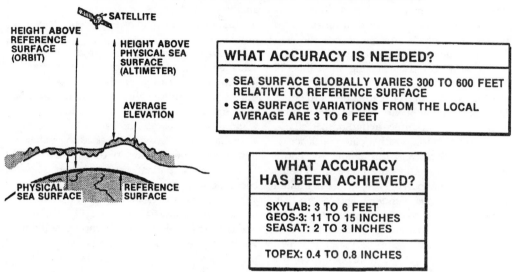

Figure 7.4 The Topex oceanographic satellite is being designed to measure the undulating surface contours of the oceans using dual-frequency radar beams transmitted vertically downward toward the earth. A GPS receiver called the *Monarch* will be carried aboard the TOPEX satellite to help fix its orbital position using precise carrier-aided interferometry techniques.

of the world approximately once every seven years.[1] El Nino is triggered by a large, periodic depression in the South Pacific off the coast of Peru.

It alters gross weather patterns on a global scale, dropping excessive rainfall in some countries, creating bone-dry droughts in other, unfortunate, parts of the world. Meteorologists are convinced that, if they can measure the oceanic depression from orbiting satellites over an extended period, they may eventually learn how to predict major weather trends within continent-sized areas a month or more in advance. When it comes, El Nino profoundly impacts the global economy. Peru's anchovy catch sometimes declines by 80 percent due to the oceanic disturbances it brings. Anchovy shortages, in turn, spark sharp increases in the value of soy beans, which are also widely used in animal feed.

How can the TOPEX oceanographic satellite manage to measure ocean surface contours with such an awesome degree of precision? Everything about its mission is designed to achieve precise and accurate ranging solutions. The satellite will be boosted into a low-drag orbit, 750 nautical miles above the earth. That particular orbital location was selected because it gives the satellite a 10-day repeating ground trace so that it will pass over the same geographical regions repeatedly to provide highly accurate calibrations.

---

[1]El Nino means "The Christ Child." It gets its name from the fact that the bad weather it brings often tends to occur near the annual holiday season.

Taking Anchovy Inventories from Outer Space

Bone-chilling rains sweep across Florida's orange groves. Hail in Kansas devastates winter wheat. Prolonged drought along the San Joaquin Valley wilts the iceberg lettuce. What in the world could be causing all of this? El Nino!

During the past decade, meteorologists have learned that El Nino—a periodic depression on the ocean surface in the South Pacific—strongly influences weather patterns all across the Pacific Basin and throughout the entire world. El Nino centers itself on a small patch of ocean off the coast of Peru, where 13 million tons of anchovies, slender finger-length fish, are harvested in an average year. Only a few Peruvian anchovies are consumed directly by enthusiastic pizza eaters, but, all together, those little fishlings constitute one-fifth of the world's annual catch of commercial fish. Unfortunately, when El Nino slices across the Pacific every seven years or so, anchovy harvests plummet, soybean futures rise, and the cost of anchovy-fed chickens in the small towns of South Carolina shoot up by 40 percent. Or even more.

Peruvian fisheries are so productive because cold water gushes toward the surface carrying a rich load of nutrients to feed tiny microscopic plants. El Nino disrupts these delicate upswelling patterns by sloshing warm water eastward across the Pacific and southward along the Peruvian coast, idling fishermen and starving millions of coastal birds.

Fortunately, patient meteorologists have been training themselves to read El Nino's subtle footprints. "The region forms the core of the heat engine that drives the world's weather and climate," explains oceanographer William Patzert of the Scripps Institute of Oceanography. Puzzle out El Nino and everything else about the weather may quietly fall into place for beleagured meteorologists searching for hidden clues on the powerful forces that shape worldwide weather trends.

Precise laser ranging devices at Bermuda will be used to measure the Keplerian orbital elements of the TOPEX satellite, dual-frequency altimeter transmissions will correct for ionospheric delays, and interferometry techniques using signals from the GPS satellites will position the TOPEX to within approximately 4 inches in space. Fortunately, most of its 4-inch positioning error will be "along track" (in the direction of travel of the space-craft), so the corresponding ranging error to the ocean surface will be much smaller. This is true because the range being measured is roughly perpendicular to the along-track positioning error.

The TOPEX oceanographic satellite being jointly sponsored by NASA's Jet Propulsion Laboratory and the French government will provide much of the vital information needed to pinpoint the next El Nino as it begins to form. Positioned in orbit by the GPS carrier waves, TOPEX will carry a sensitive radar altimeter capable of measuring the heights of the oceans to an almost unbelievable accuracy—somewhere between three-eighths and three-quarters of an inch! It will orbit in a 750-nautical-mile polar orbit that carries it around the earth every 104 minutes.

Pinning down El Nino will likely help the people of Peru, but it will also be of benefit to many others living in distant lands. El Nino's curious predictability has helped convince today's weather experts that, if they focus their attention on interactions between the earth's many forms of energy— sun, waves, tides—they may eventually learn how to predict average weather patterns days, weeks, even months in advance.

Motorola's Commercially Available Monarch

The spaceborne carrier-aided receiver designed for use aboard the TOPEX oceanographic satellite is being advertised by Motorola under the trade name *Monarch*. The Monarch gets its name from the colorful butterfly that migrates over vast distances with such impressive precision.

Motorola's Monarch is a 2-channel receiver fully qualified for successful operation in outer space. It uses pseudo-range measurements to obtain its first-guess position, and then sharpens that initial estimate with carrier-aided computer processing techniques.

Tomorrow's Generic Spaceborne Receivers

Several teams of engineers have been examining various possibilities for more precise and accurate spaceborne receivers. One of them, the Space Navigation and Pointing System (SNAPS) has been proposed by researchers

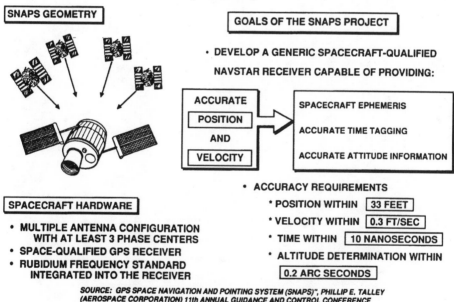

GPS SPACE NAVIGATION AND POINTING SYSTEM (SNAPS)

SNAPS GEOMETRY

GOALS OF THE SNAPS PROJECT

• DEVELOP A GENERIC SPACECRAFT-QUALIFIED NAVSTAR RECEIVER CAPABLE OF PROVIDING:

ACCURATE POSITION AND VELOCITY → SPACECRAFT EPHEMERIS / ACCURATE TIME TAGGING / ACCURATE ATTITUDE INFORMATION

SPACECRAFT HARDWARE

• MULTIPLE ANTENNA CONFIGURATION WITH AT LEAST 3 PHASE CENTERS
• SPACE-QUALIFIED GPS RECEIVER
• RUBIDIUM FREQUENCY STANDARD INTEGRATED INTO THE RECEIVER

• ACCURACY REQUIREMENTS
  * POSITION WITHIN   33 FEET
  * VELOCITY WITHIN   0.3 FT/SEC
  * TIME WITHIN   10 NANOSECONDS
  * ALTITUDE DETERMINATION WITHIN   0.2 ARC SECONDS

SOURCE: GPS SPACE NAVIGATION AND POINTING SYSTEM (SNAPS)", PHILLIP E. TALLEY (AEROSPACE CORPORATION) 11th ANNUAL GUIDANCE AND CONTROL CONFERENCE AMERICAN AERONAUTICAL SOCIETY, KEYSTONE, COLORADO, JAN. 30-FEB. 3, 1988.

Figure 7.5 Researchers at the Aerospace Corporation have proposed the development of this highly capable spaceborne receiver that would ride into space aboard various satellites to provide accurate real-time measurements of position, velocity, spacecraft attitude, and the exact time. Equipped with an onboard rubidium atomic clock and three or four carefully constructed antennas, the attitude measuring portion of this 15-pound device might be able to fix the satellite's instantaneous orientation to within 0.2 arc seconds, an accuracy that is fully competitive with most of today's spaceborne inertial navigation systems.

at the Aerospace Corporation in El Segundo, California. The SNAPS space-borne receiver (see Figure 7.5) is slated to fix its position, obtain highly accurate time, and determine its orientation angles using precise interferometry techniques.

The SNAPS unit will be rigged with three or four antenna phase centers. A triangular configuration with three phase centers is being seriously considered so is a pyramid configuration with four phase centers in three-dimensional space. A rubidium atomic clock carried onboard the spacecraft will provide precise time and frequency measurements to help smooth and stabilize the interferometry solutions.

Proponents of the SNAPS concept are convinced that its errors may be as small as 30 feet in position, 0.3 feet-per-second in velocity, and 10 nanoseconds in time. Attitude angles accurate to within 0.2 arc seconds are also envisioned by the system engineers. Total target weight for the highly compact system is only 15 pounds. If it can be developed and successfully space-qualified, the multipurpose SNAPS will undoubtedly find a multitude of applications in tomorrow's military and commercial arenas deep in outer space.

8

Integrated Navigation Systems

In 1927, when Charles Lindbergh ventured out over the Atlantic aboard his beloved "Spirit of St. Louis," he tracked his progress using dead reckoning positioning techniques. As accurately as he could, Lindbergh measured his ground speed along each leg of his 33-hour journey. Then he multiplied by the elapsed time to estimate his new position.

Under ideal conditions, dead reckoning fixes the user's position with reasonable precision, but it is a cumulative process, so errors build up with the passage of time. If "Lucky Lindy" could establish his position within 5 miles as he passed over Newfoundland, then 20 hours later, as he knifed across the French coastline, his position uncertainty might then equal 30 miles or more. Of course, if he could recognize familiar landmarks along the way, his position uncertainty might shrink to a mile or so, before it began to build back up again.

Most modern navigation devices employ triangularization techniques with external signals to fix position, so, as time passes, their errors do not tend to accumulate. But in keeping with the exciting traditions established by Charles Lindbergh, one important exception still exists: the inertial navigation system.

The earliest inertial navigation systems were fashioned from enlarged models of toy gyroscopes rigged to carry spring-actuated devices called accelerometers. By measuring its current acceleration increments then multiplying by the elapsed time, an inertial navigation system determines its instantaneous velocity. By keeping track of its velocity and multiplying by the appropriate increment of time, it determines three-dimensional change in its position.

Position uncertainties gradually accumulate, but, by recognizing familiar landmarks along the way—navigation transmitters, for instance, or known stars at precise locations—an inertial navigation system can periodically shrink its position uncertainty to a much smaller magnitude. Charles Lindbergh died many years ago. But if he was still alive, he would be much impressed with today's modern refinements in dead reckoning position-fixing techniques.

Integrated Navigation

One popular type of integrated navigation system links an inertial navigation unit with a GPS receiver so that both devices can contribute inputs to the current navigation solution. A combined system of this type represents a beautiful marriage of modern technology, and it provides consistently superior results compared with those that could be achieved by either unit working alone.

Integrated navigation systems have been constructed, tested, and installed aboard many different types of vehicles, including helicopters, airplanes, surface ships, and submarines. Expensive and sophisticated components have often been utilized, but many knowledgeable experts maintain that the most beneficial marriages often involve inexpensive inertial navigation systems linked with inexpensive Navstar receivers.

Compared with a stand-alone GPS receiver, an integrated navigation system usually provides substantial improvements in accuracy and stability, jamming immunity, and high-dynamic operation. The major shortcoming of an integrated navigation system centers around the extra cost associated with purchasing and merging the two separate units and the extra cost and complexity of their combined Kalman filtering routines.

Inertial Navigation

An inertial navigation system is a simple, self-contained position-fixing device that continuously measures three mutually orthogonal acceleration components, numerically integrates those accelerations to obtain the instantaneous velocity, and then integrates the resulting velocity to determine the vehicle's current position.

Two fundamentally different types of inertial navigation systems are currently in widespread use:

1. Gimballing inertial navigation systems
2. Strapdown inertial navigation systems

Initially, gimballing systems were by far the most popular, but in recent years strapdown systems have captured an ever-increasing market share.

In a *gimballing inertial navigation system* the three mutually orthogonal accelerometers are rigidly mounted on the inner gimbal of three nested gyroscopes. As the parent craft executes maneuvers of any desired complexity, the inner gimbal always maintains a constant attitude. Its attitude may be fixed in inertial space (see Figure 8.1), or it may be parallel to the local vertical orientation. A local vertical device uses a feedback control loop based on the "Schuler Pendulum" to maintain its earth-seeking orientation.[1]

In a *strapdown inertial navigation system* the three mutually orthogonal accelerometers are mounted parallel to the body axes of the parent craft. Changes in vehicle attitude are continuously measured by using gyroscopic sensing techniques, but the accelerometers themselves do not twist and turn as the vehicle executes its maneuvers. A strapdown system is physically and conceptually simpler than a gimballing system, but is involves much larger computational burdens. With modern microprocessor circuits, this is not a very serious problem.

Inertial navigation systems are blessed with a number of advantageous characteristics. In particular, they are small, compact, self-contained, and highly survivable. Generally speaking, the only practical way to destroy an inertial navigation system is to destroy the vehicle on which it is installed. A radionavigation system, by contrast, can be knocked out by destroying its transmitters or, in some cases, by destroying the control facilities used in updating those transmitters.

THE 3-AXIS STABLE PLATFORM

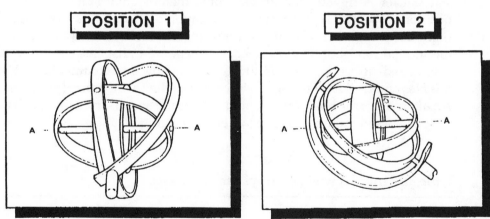

| POSITION 1 | POSITION 2 |

Figure 8.1 A gimballing inertial navigation system usually consists of three gyroscopically stabilized platforms triply nested one inside the other. Three mutually orthogonal accelerometers mounted on the innermost gimbal measure the instantaneous acceleration vector. Computerized numerical integration procedures are then used to determine the vehicle's current velocity. A second integration yields its position.

---

[1]A Schuler pendulum oscillates with a period of 84 minutes. A simple pendulum with this period would be 4,000 miles long. Cleverly constructed feedback control loops shorten its length to a more appropriate magnitude of only a few inches.

Error Growth Rates

An inertial navigation system uses dead-reckoning techniques to keep track of its position, so its positioning errors tend to grow relentlessly with time. A medium accuracy system, for instance, might have an error growth rate of 1 nautical mile per hour. This means that within an hour after it is initialized, it will build up a 1 nautical mile uncertainty in its position, an hour later it will have accumulated a 2 nautical mile uncertainty, and so on.

Some high-accuracy inertial navigation systems can be constructed with error growth rates ranging from 0.1 to 0.25 nautical miles per hour, but they are extremely costly and complicated. The values in Table 8.1 define the error growth profiles for medium-accuracy and high-accuracy inertial navigation systems. The equation running across the bottom of the table further defines the error characteristics for systems of this type.

Reinitialization Techniques

Some inertial navigation systems, including the ones carried aboard short-range missiles, operate as stand-alone units. But most use external sources to limit their growing errors within reasonable bands. Star trackers, for instance, are often used to update inertial navigation systems. An onboard star tracker is rigged with an electronic memory that stores the relative positions of several highly visible stars scattered across the sky. During each flight the star tracker takes precise signtings on selected stars to reinitialize the inertial navigation system. Star trackers are indispensable for certain specialized missions, such as those handled by intercontinental ballistic missiles and the Hubble Space Telescope, but they are extremely expensive. An alternate error-bounding technique uses the signals from ground based or space-based radionavigation systems for reinitialization.

Table 8.1   Inertial navigation system accuracies

Type Of INS	Position error (CEP)	Velocity error (RMS)
Medium accuracy	1 n.mi./hr	2.5 ft/sec
High accuracy	0.25 n.mi./hr	1.5 ft/sec

Error Characteristics for an inertial navigation system:
$P(t) = a + bt + c \sin \sigma_s t + \phi) + d\sqrt{t}$
  a = initial bias error
  bt = initial error growth rate
  $c \sin \sigma_s t + \phi$ = Schuler periodic errors
  $d\sqrt{t}$ = random walk errors

A more effective, but more costly, approach consists of integrating the inertial navigation device with a radionavigation system to form a single unit so that the two devices can mutually influence each current navigation solution. Integrated navigation systems will be discussed in a later section, but first ring-laser and fiber-optic gyros must be introduced.

Ring Laser Gyros

A ring laser gyro is, to some extent, conceptually similar to a mechanical strapdown inertial navigation system, but it does not use mechanical sensing techniques to determine current attitude angles. Instead, it uses interferometry techniques to sense the rotation rates of the parent craft.

As shown in Figure 8.2, two counter-rotating laser beams are reflected around a hollow cavity inside a ring laser gyro. If the cavity is rotating in the clockwise direction, the clockwise carrier waves will be stretched out while the counterclockwise carrier waves will be compressed. Thus, when

RING LASER GYRO OPERATION

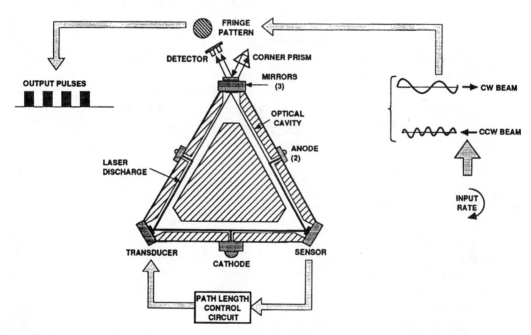

Figure 8.2 Counterrotating laser beams are reflected around a hollow cavity inside a ring laser gyro. If the device is rotating in the clockwise direction, the clockwise carrier waves will be stretched out as they travel around the circuit. The counterclockwise carrier waves will be compressed. When they come back together, the two waves are beat against one another to create interference fringe patterns which sweep by an optical detector. This allows the unit to measure its current rotation rate.

the two laser beams come back together, at their point of origin, they will be slightly out of phase. Beating the two counter-rotating waves together creates interference fringe patterns consisting of alternate bands of dark and light. By allowing these interference fringe patterns to sweep by an optical detector, the rate at which they are being created can be measured. This gives a direct indication of the unit's current rotation rate. Three mutually perpendicular hollow cavities, each served by its own dedicated laser beam, can be linked together to completely define any twisting and turning of the parent craft.

In general, the rate at which the dark and light bands sweep by the optical detector is directly proportional to the rotation rate. But, when the rate of rotation approaches zero, the device enters a so-called "dead band," in which no interference fringe patterns are produced. Fortunately, "dithering" techniques can compensate for this dead band phenomenon. A ring laser gyro can be dithered by using piezoelectric transducers to rock its mirrors back and forth at a high rate of speed. This produces a series of rotation rate readings on alternate sides of the dead band that can be averaged to detect and measure even the smallest rates of rotation.

A functioning inertial navigation system usually requires three hollow cavities arranged in a mutually orthogonal configuration to measure the three separate rotation rates. When these three rates are coupled with the outputs from three integrating accelerometers, the system can be used to determine the parent craft's current velocity and position.

All other things being equal, a ring laser gyro with a longer perimeter will yield improved positioning accuracy. Of course, a longer perimeter generally corresponds to a larger and heavier device, which is harder to fit into a ship or plane. For this reason, monolithic ring laser gyros have been developed. They are smaller, lighter devices with ample perimeters. Monolithic ring laser gyros are also simpler, smaller, and cheaper than ring laser gyros with more conventional architectures.

Monolithic Ring Laser Gyros

The sketch in Figure 8.3 highlights the geometrical features of a typical monolithic ring laser gyro. Notice that its unique architecture provides three *square* laser beam paths, using a total of only six mirrors. Three mutually orthogonal square paths are possible with such a small number of mirrors because each mirror is used twice; it is positioned along two of the three square paths. Conceptually, the mirrors can be regarded as being mounted on the faces of a cube. This clever geometrical arrangement reduces the parts count for the unit, simplifies its overall complexity, and cuts its manufacturing and alignment costs. In addition, long path lengths are provided with a highly compact design.

A monolithic ring laser gyro typically weighs only about 40 percent as much as a ring laser gyro of conventional design. Its improved architecture

IMPROVED ARCHITECTURE FOR THE SECOND GENERATION RING LASER GYRO's

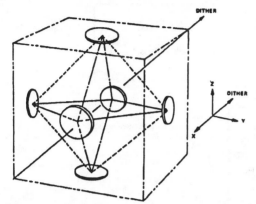

THE GEOMETRY

- **NESTED OPTICAL CAVITIES WITH SIX MIRRORS MOUNTED ON THE FACES OF A CUBE (CONCEPTUAL GEOMETRY)**

- **THIS GIVES 3 SQUARE ORTHOGONAL GYROS WITH EACH MIRROR SHARED BY 2 OF THE PATHS**

Figure 8.3   The monolithic ring laser gyro employs three square laser beam paths, using only six mirrors. The mirrors are mounted on the faces of a cube, so each one lies along two of the three rectangular paths. With its novel architecture, the monolithic ring laser gyro has longer path lengths and simpler design features than most conventional ring laser gyros.

also yields cleaner mechanical interfaces and simpler dithering techniques. However, an even simpler design approach, the fiber optic gyro, is being perfected by industry experts.

Fiber Optic Gyros

In a fiber optic gyro two counter-rotating laser beams are routed in opposite directions through a hair-thin optical fiber tightly wound around a cylindrical spool. As with the ring laser gyro, interference fringe patterns are created at a rate that is directly proportional to the rotation rate of the parent craft.

Path lengths of 3,000 feet are commonly used, so it may seem that a fiber optic gyro would yield incredibly accurate navigation. Unfortunately, other design difficulties, such as centering the laser beam within the fiber, tend to degrade its accuracy. For this reason, fiber optic gyros have, so far, been used primarily for low-accuracy applications in which small size, low weight, and rugged construction are at a premium.

As Figure 8.4 shows, extremely small fiber optic gyros can be constructed using relatively simple fabrication techniques. Litton Industries has fashioned a functioning unit weighing only about a quarter of a pound. It features three mutually orthogonal spools wound with optical fibers served by a triad of tiny solid-state accelerometers.

The accelerometers are machined from solid monolithic slabs of silicon using mass production techniques perfected in the digital watchmaking industry. The tiny silicon devices feature vibrating "strings" of silicon whose

FIBER-OPTIC GYROS FOR THE 1990's

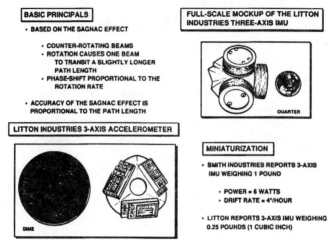

BASIC PRINCIPALS
- BASED ON THE SAGNAC EFFECT
  - COUNTER-ROTATING BEAMS
  - ROTATION CAUSES ONE BEAM TO TRANSIT A SLIGHTLY LONGER PATH LENGTH
  - PHASE-SHIFT PROPORTIONAL TO THE ROTATION RATE
- ACCURACY OF THE SAGNAC EFFECT IS PROPORTIONAL TO THE PATH LENGTH

LITTON INDUSTRIES 3-AXIS ACCELEROMETER

DIME

FULL-SCALE MOCKUP OF THE LITTON INDUSTRIES THREE-AXIS IMU

QUARTER

MINIATURIZATION
- SMITH INDUSTRIES REPORTS 3-AXIS IMU WEIGHING 1 POUND
  - POWER = 6 WATTS
  - DRIFT RATE = 4°/HOUR
- LITTON REPORTS 3-AXIS IMU WEIGHING 0.25 POUNDS (1 CUBIC INCH)

Figure 8.4   The counterrotating laser beams in a fiber optic gyro travel through optical fibers wound around three cylindrical spools. Long path lengths are thus provided with a highly compact design. Fiber optic gyros are not yet competitive with ring laser gyros in terms of accuracy. But they are simple, small, and cheap. Some commercially available units weigh less than four ounces each.

vibration rates vary in response to the acceleration forces acting on the parent craft. Some of today's solid-state accelerometers are smaller than a baby's fingernail.

Using the GPS for Testing Inertial Navigation Systems

Once a navigation system has been designed and produced, some means of proving that it meets performance specifications must be devised. Recently, a division of Litton Industries at Moorpark, California, demonstrated a cleverly conceived certification technique in which company engineers harnessed the unique navigation capabilities of the Navstar GPS in an attempt to prove that a new inertial navigation system would work as advertised.

For their certification tests, the Litton engineers mounted the new monolithic ring laser gyro together with a company-built GPS receiver on a specially equipped truck, which they drove along the freeways in San Fernando Valley. As the truck traveled on its prearranged route, they compared the positioning solutions from the two different navigation systems to see if the ring laser gyro's performance specifications were being achieved.

Laboratory analyses of the resulting position and velocity data clearly revealed that, as expected, four different types of error components were being observed:

- Initial bias errors
- Initial growth-rate errors
- Schuler periodic errors (with an 84-minute period)
- Random walk errors

This highly efficient method of testing will surely be adopted by other manufacturers, but the most promising future for the joint use of an inertial navigation system and the GPS definitely resides with integrated navigation, in which the two units are made to act in close partnership to improve their mutual performance capabilities.

The Practical Benefits of Integrated Navigation

Mate an inertial navigation system with a GPS receiver and the new couple can almost always make beautiful music together. As indicated in Figure 8.5,

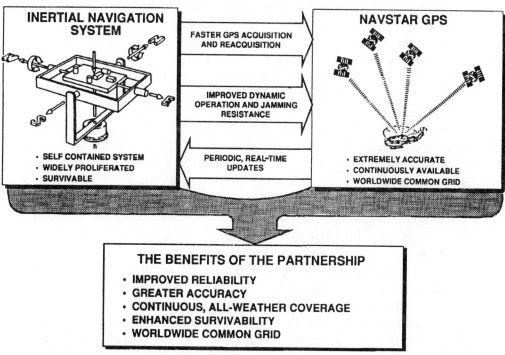

HOW INERTIAL NAVIGATION AND GPS CAN HELP ONE ANOTHER

INERTIAL NAVIGATION SYSTEM

NAVSTAR GPS

FASTER GPS ACQUISITION AND REACQUISITION

IMPROVED DYNAMIC OPERATION AND JAMMING RESISTANCE

PERIODIC, REAL-TIME UPDATES

- SELF CONTAINED SYSTEM
- WIDELY PROLIFERATED
- SURVIVABLE

- EXTREMELY ACCURATE
- CONTINUOUSLY AVAILABLE
- WORLDWIDE COMMON GRID

THE BENEFITS OF THE PARTNERSHIP

- IMPROVED RELIABILITY
- GREATER ACCURACY
- CONTINUOUS, ALL-WEATHER COVERAGE
- ENHANCED SURVIVABILITY
- WORLDWIDE COMMON GRID

Figure 8.5 An integrated navigation system consists of an inertial navigation system working in partnership with a GPS receiver. It provides better reliability, greater accuracy, and enhanced survivability compared with conventional stand-alone units. The GPS receiver helps the inertial navigation system by providing faster signal acquisition and reacquisition whenever loss of lock occurs. It also improves the jamming immunity and dynamic operation of the GPS receiver. The GPS receiver helps the inertial navigation system by providing it with real-time information on the current behavior of its error statistics.

better reliability, smaller navigation errors, and improved survivability can all be listed among the practical benefits that will likely result.

In particular, the inertial navigation system can help the GPS by providing accurate initial estimates on position and velocity, thus reducing the time required for the GPS receiver to lock onto the signals streaming down from the satellites. If one or more of the satellite signals is subsequently lost due to receiver malfunction, terrestrial blockages, wing shielding, or enemy jamming, the inertial navigation system can help achieve reacquisition quickly and efficiently. The continuous velocity measurements obtained from the inertial system allow the GPS receiver to estimate the magnitude of the current Doppler shift so that it can narrow the bandwidth of its tracking loops. This improves the dynamic operation of the integrated navigation system and also increases its jamming immunity.

The GPS receiver can help the inertial navigation system with accurate, real-time estimates on the current behavior of its error statistics. This helps it to provide more stable and accurate navigation solutions.

Many industry experts believe that the most beneficial results can be achieved by using some of the cheapest available hardware units when building an integrated navigation system. This concept was verified to some extent when Northrup engineers mated a single-channel GPS receiver with an inertial navigation system rigged with low-cost ring laser gyros and low-cost accelerometers. The performance of the resulting integrated navigation system turned out to be roughly equivalent to an expensive 5-channel GPS receiver, in terms of jamming immunity and dynamic operating capabilities. The mating also produced many other advantageous design characteristics.

The graphs in Figure 8.6 compare the performance capabilities of a 5-channel GPS receiver with its performance when it has been integrated with an inertial navigation system. Actual production equipment was used in these tests, but the signals from the GPS satellites were simulated by computer. This unique arrangement allowed the test engineers to model the system response using various satellite constellations, signal strengths, and equipment combinations. Aircraft trajectories of various types were also simulated during the course of the study.

The graphs in Figure 8.6 correspond to a case in which the aircraft is traveling around an elongated "race course" trajectory. On each turn it tilts at a 60-degree bank angle so that its wings block the L-band signals from two of the four Navstar satellites. Each time two of the satellite signals are lost, tall spikes appear in the navigation error. The worst-case errors amount to about 230 feet. Of course, when the airplane levels off again, the error spikes disappear.

The dotted line represents a different case in which the Navstar receiver is integrated with an inertial navigation system. In this case, both units are influencing the positioning solution and, as a result, the error spikes completely disappear. Notice that the positioning error never rises above 50 feet, even during the blockage intervals.

5-CHANNEL AIDED VS. UNAIDED SIMULATION

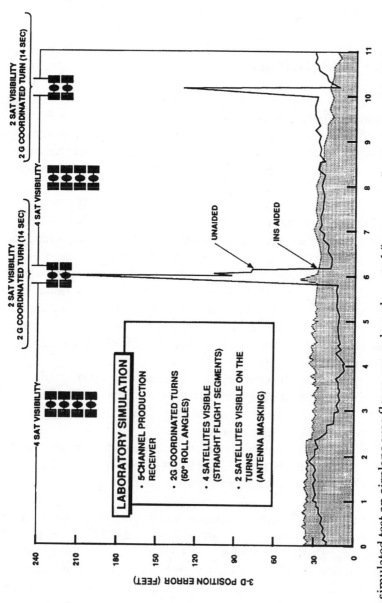

Figure 8.6 In this simulated test an airplane was flown around an elongated "race course" trajectory with 60-degree bank angles on the turns. Most of the time the GPS receiver was locked onto four Navstar satellites, but, at high bank angles, tow of the four satellite signals were blocked by the wings. When the GPS was acting alone, large error spikes occurred during the two blockage intervals, but, when the GPS receiver was integrated with an inertial navigation system, the error spikes disappeared.

113

Notice, also, that, when the integrated navigation system is positioning the aircraft, the errors turn out to be slightly larger than for the comparable solution in which the GPS receiver is acting alone. This occurs because, although the GPS is known to be more accurate than the inertial navigation system, if inertial navigation is not given some influence on the solution, the spikes that occur during the outage intervals would tend to persist even when integrated navigation is being used.

Chassis-level Integration

In an integrated navigation system, the GPS receiver and the inertial navigation system usually consist of simple, stand-alone units whose outputs are, together, fed into one or more Kalman filters. But a different approach called chassis-level integration is also being pursued by navigation industry experts. With chassis-level integration, a GPS receiver is inserted right inside an existing inertial navigation system.

This novel approach results in a combined unit with the same form, fit, and function and the same interfaces as the original inertial navigation system. Practical benefits include simpler interfaces and timing requirements and simpler communication links between the two devices.

But how can a GPS receiver be inserted inside an existing inertial navigation system? This is possible because the electronic components in today's Navstar receivers are surprisingly small and compact—playing-card size devices are passing into general use—and because a few of the components in an existing inertial system can be replaced with miniaturized versions to provide more room.

Popular miniaturization procedures include the use of gate arrays and programmable logic arrays, the introduction of advanced VHSIC[2] technology and the use of space-saving mountings, such as pin-grid arrays and compact surface-mounting techniques.

Once a GPS receiver has been successfully inserted inside the existing inertial navigation system, installation into space-limited vehicles is greatly simplified. Depending on the application, navigation performance can also be strongly enhanced.

In one test series, McDonnell-Douglas Astrionics demonstrated a reacquisition interval amounting to only about 6 seconds, following loss of lock. their test engineers also noticed a 13dB extra jamming margin compared with a comparable stand-alone unit, together with an increase in dynamic operating capabilities from 2.3 g's to 24 g's. Other researchers have observed and documented similar performance enhancements whenever GPS/INS integration has been successfully achieved.

---

[2]VHSIC = *Very High-Speed Integrated Circuits.*

9

Interoperability with Other Navigation Systems

Standing alone, the Navstar GPS provides extremely versatile and accurate position-fixing, but, if it can be combined with various other navigation systems, its versatility and accuracy can be further increased. In particular, a successful marriage between the former Soviet Union's Glonass Constellation and the American GPS could yield extremely beneficial results.

Such a combined system could provide accurate global navigation coverage much sooner than would otherwise be possible, and it may help foster onboard integrity-monitoring for air-related users. These and other promising benefits from a workable GPS/Glonass partnership have led a number of enthusiasts to encourage cooperative ventures between the technological communities in the two countries, with an eye toward constructing efficient dual-capability GPS/Glonass receivers.

The Soviet Glonass

Like their American counterparts, the Glonass satellites masterminded by the Commonwealth of Independent States are being launched into circular, inclined orbits approximately 11,000 nautical miles above the earth. And, like their American colleagues, the scientists in the Commonwealth Countries plan to orbit a constellation of 21 satellites plus 3 active on-orbit spares. The users in both systems measure the pseudo-ranges and the pseudo-range rates to four or more satellites, and then use trilateralization procedures to determine their current positions and velocities—and the exact time.

Table 9.1 highlights some of the other similarities as well as a number of differences between the two spaceborne navigation systems. The Glonass

115

Table 9.1 Comparisons between the American GPS and the Soviet Glonass

	American GPS	Soviet Glonass
Planned constellation	21 1 3 active on-orbit spares	21 1 3 active on-orbit spares
Number of orbital rings	6	3
Orbital altitude	10,898 n. mi.	10,313 n. mi.
Orbital inclination	63° (Block Is) 55° (Block IIs)	64.8°
Orbital period	12 hours	11 hours 15 minutes
Repeating ground- trace interval	1 day (2 orbits)	8 days (17 orbits)
First launch	Feb 1978	Oct 1982
Booster rockets	Atlas F, Delta II	SL-12 Proton
Launch sites	Vandenberg, Cape Kennedy	Tyuratam
Signal access method	Code-Division Multiple Access	Frequency-Division Multiple Access
C/A-code repetition interval	1,023 bits	511 bits
C/A-code bandwidth	2 megahertz	1 megahertz
Data Stream Bit Rate	50 bits/second	50 bits/second
L-band polarization	Right-hand circular polarized	Right-hand circular polarized
Civilian positioning accuracy (CEP) (horizontal plane)	330 feet	330 feet

satellites are being launched into three orbital rings, for instance, as compared with six rings for the GPS. And the Glonass transmitters use frequency-division multiple access to distinguish among the various satellites, rather than code-division multiple access techniques. In other words, each Glonass satellite is assigned its own unique frequency to distinguish it from its neighbors in space.

The Glonass satellites are being launched into 10,313–nautical mile orbits, with orbital periods of 11 hours and 15 minutes. Their ground-trace repetition interval is 17 orbits (8 days), as compared with the GPS orbits, which repeat after only two orbits (1 day).

The Glonass scientists are launching their satellites into 64.8-degree orbits, as compared with 55 degrees for the Block II GPS satellites. Their C/A-code repetition interval is 511 bits, as compared with 1023 bits for the Navstar satellites, and their C/A-code bandwidth is 1 megahertz, compared with 2 megahertz for the similar GPS C/A-code transmissions. The corresponding

P-code transmission bandwidths are 10 and 20 megahertz, respectively, for the Russian and the American satellites.

The Glonass Specification Release at Montreal

In 1989, at the Future Air Navigation Systems (FANS) meeting in Montreal, navigation experts from the former Soviet Union released the nonmilitary specifications for their Glonass constellation. The Soviet experts attending that meeting indicated that they were hoping to attract a worldwide class of civilian users to begin accessing their Glonass navigation signals for a broad range of nonmilitary applications. The Soviet scientists estimated that the positioning error for Glonass navigation would be 330 feet or less in the horizontal plane. The advertised velocity error for users of their system was set at 0.5 feet per second. These quoted values are comparable to the corresponding degraded civilian errors for Navstar navigation.

The Glonass Constellation

Like America's Navstar GPS, the Glonass constellation has already been partially installed in space. At least 46 Glonass satellites have been hurled into orbit aboard SL-12 Proton boosters from the heavily used launch complex at Tyuratam. Each SL-12 carries three Glonass satellites into a 10,313–nautical mile orbit tipped 64.8 degrees with respect to the equator.

According to reports published in the open literature, the Glonass constellation is scheduled to reach full operational status by the middle of the 1990s. Sketches of a typical Glonass satellite and a typical Glonass receiver are presented in Figure 9.1. Notice that the satellite is constructed with a largely cylindrical center-body section. Its axial symmetry may be helpful in minimizing the variations in solar radiation pressure that inevitably result when different portions of the center body are oriented toward the sun.

The architecture of America's 6-plane GPS constellation was purposely selected so that pairs of GPS satellites would never pass close together. This enhances the survivability of the constellation because it forces any enemy to engage the GPS satellites one-on-one. For some reason, the satellites in the Glonass constellation pass together in pairs at 47 degrees North latitude and 47 degrees South latitude, as shown in Figure 9.2. For each pair of coincident satellites, these close encounters are repeated every three hours.

Orbital Maneuvers for the Glonass Satellites

Aerospace engineers from the Commonwealth of Independent States have achieved superb insertion accuracies in launching their Glonass satellites into the desired 10,313–nautical mile orbits. If a satellite's orbit is perfectly

SOVIET GLONASS HARDWARE

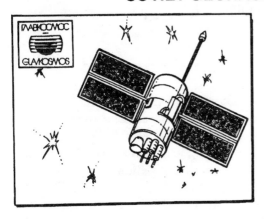

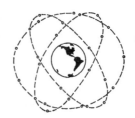

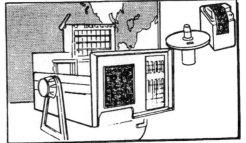

SOURCE:
"NAVIGATIONAL ASPECTS OF GLONASS "
MOSKUIN (USSR MINISTRY OF MERCHANT MARINE
AND SOROCHINSKY, CENTRAL MARINE RESEARCH
INSTITUTE) GPS WORLD

Figure 9.1   The Soviet Glonass satellite (upper left) features a largely cylindrical center-body section sprouting two winglike solar arrays, three circular polarized navigation antennas, and a spear-shaped tracking and telemetry antenna. The box-like Glonass receiver (bottom right) picks up the L-band signals from the Glonass satellites circling overhead in three orbit planes. Industry experts estimate that a typical Glonass receiver is about four times as big and heavy as its GPS counterpart. However, the Soviets are rumored to be building a much smaller hand-held receiver.

circular, its apogee altitude and its perigee altitude will both be the same. For the Glonass missions the apogee and perigee altitudes never seem to differ by more than about 60 nautical miles. This compares with an insertion error of nearly 300 nautical miles for the least accurate orbits of the American GPS.

Once they reach their final destination orbits, the Glonass satellites appear to have ample amounts of propellant for large and frequent on-orbit maneuvers. In 1981, when the atomic clocks onboard America's Navstar 1 failed, Navstar 2 was moved 180 degrees within the orbital ring to take its place. Only a few pounds of hydrazene monopropellant were available to power this slow, lumbering maneuver, and, as a result, it took 6 months to reposition Navstar 2. By contrast, in September 1988, when the Soviet scientists repositioned one of their Glonass satellites (COSMOS 1885), only 10 days were required to move it 90 degrees. Another Glonass satellite (COSMOS 1779) was later moved at a comparable rate 45 degrees on three separate occasions.

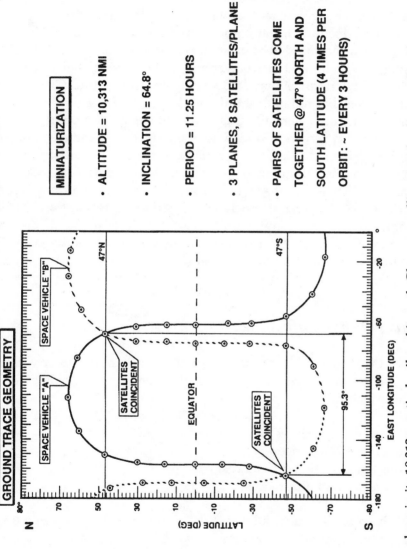

Figure 9.2 Coasting along in its 10,313-nautical mile orbit, each Glonass satellite completely encircles the earth every 11 hours and 15 minutes. Thus, it returns to the same spot over planet Earth when it has completed 17 orbits (8 days). Most of the time the Glonass satellites are hundreds or thousands of miles apart. But, for some reason, pairs of them pass into close proximity at 47-degrees North latitude or 47-degrees South latitude. Such an encounter between adjacent pairs of satellites happens every three hours.

Building Dual-capability GPS/Glonass Receivers

A number of different organizations and combinations of organizations have been attempting to build dual-capability receivers capable of picking up and processing the navigation signals from both GPS and Glonass. If desired, a dual-capability receiver could employ an overdetermined solution, with some of the navigation signals coming from GPS satellites and some coming from the Glonass satellites. This approach would increase the accuracy, reliability, robustness, and integrity of the dual-capability receiver.

The earliest documented attempt to build a dual-capability GPS/Glonass receiver was masterminded by professors Daly and Dale at Leeds University in England. With limited manpower and limited funds, they quickly managed to build a receiver that could navigate using either GPS signals or Glonass signals, but not both simultaneously.

Other organizations are also attempting to design and build other types of dual-capability receivers, often with the cooperation and encouragement of Soviet engineers. Ashtec, Inc., for example, is working with Space Device Engineering in the Commonwealth to build a dual-capability receiver, and researchers at The Federal Aviation Administration are working with Lincoln Laboratories at The Massachusetts Institute of Technology (M.I.T.) in an attempt to construct a similar unit. Experts at The Magnavox Corporation in Torrance, California, are also developing a dual-capability receiver that is said to be working to a limited extent.

Receiver Design Difficulties

Some of the difficulties that international engineers encounter when they attempt to design and build dual-capability receivers are tabulated in Table 9.2. These include:

- Time scale differences
- Ephemeris differences
- Coordinate system differences
- Access method differences

GPS System Time is related to UTC (Universal Time Coordinated), a widely accepted time standard maintained in Paris, France. The Glonass satellites, unfortunately, use an entirely different timing standard called Glonass System Time, which is related to Moscow Standard Time. Any expert who wants to build and operate a dual-capability receiver must somehow resolve the offsets and incompatibilities to achieve time synchronization between the two sets of pseudo-range measurements. This can be extremely difficult unless the users can gain access to some kind of external timing adjustment.

Table 9.2 Incompatibilities that cause problems for dual-capability GPS/Glonass receivers

Incompatibility	GPS system approach	Glonass system approach
Different time standards	GPS system time related to UTC time	Glonass system time related to Moscow time
Different ephemeris constants	Expanded set of Keplerian orbital elements	Rectilinear position coordinates, velocity components, and acceleration components
Different coordinate systems	WGS-84 oblate spheroid	SGS-85 oblate spheroid
Different access methods	Code-division multiple access	Frequency-division multiple access

The ephemeris constants used by the two navigation systems are also intrinsically different. The GPS satellites transmit an expanded version of the six Keplerian orbital elements: a, e, i, V, v, and t. The other ten ephemeris constants in the GPS data stream include the derivatives of selected Keplerian orbital elements, together with a set of correction terms defining the perturbative influences of the higher-order shape terms (potential harmonics) of the earth.

The Glonass satellites, on the other hand, transmit six rectilinear position and velocity components—x, y, z, $\dot{x}$, $\dot{y}$, $\dot{z}$—plus acceleration components. Complicated software routines are required to bring the two different types of ephemeris coordinates into precise correspondence. The enlarged set of Keplerian orbital elements transmitted by the GPS satellites can be conveniently manipulated using simple algebraic and trigonometric relationships. The mutually orthogonal components used by the satellites in the Glonass Constellation lead, more naturally, to numerical integration procedures, such as 4th order Runge-Kutta integration.

The GPS ephemeris constants are directly related to the WGS-84 longitude-latitude coordinate system, whereas the Glonass constants are related to the SGS-85 coordinates. These coordinate incompatibilities add additional software routines and computational burdens to any mutually compatible receiver.

The transmission frequencies for the two navigation systems are also different, and so are the methods for accessing the navigation codes they broadcast. The Navstar Global Positioning System relies on code-division multiple access techniques. This means that all 24 GPS satellites operate on the same two transmission frequencies, L_1 and L_2. The L_1 and L_2 signals coming down from the various GPS satellites are distinguished solely by the fact that each satellite is assigned its own unique C/A-code and its own unique P-code.

By contrast, the Glonass satellites employ frequency-division multiple access techniques. Each Glonass satellite broadcasts its navigation signals on its own unique frequency to allow the various users to distinguish it from the other satellites in the Glonass constellation. Because the two systems employ different transmission frequencies and different access methods, users must employ two separate antennas—or construct a special antenna capable of picking up both types of signals simultaneously. Manufacturers would prefer to create a single dual-capability antenna, but it is not yet clear how to design and build it.

Dual-Capability Receiver Tests at Leeds University

The first dual-capability receiver built by Daly and Dale at Leeds University was equipped with a high-quality rubidium atomic clock to help maintain proper time synchronization. Its software routines totaled 15,000 lines of code, each line of which had to be planned, written, and carefully debugged. This compares with 60,000 lines of code that were written for the 5-channel military receivers developed by Rockwell International for the DoD. The Leeds University receiver determines the current orbital position of the Glonass satellites using 4th order Runge-Kutta predictor-corrector numerical integration, with a 60-second integration step size.

Tests conducted on the campus at Leeds have yielded acceptable, but not spectacular, navigation accuracies for the dual-capability GPS/Glonass receiver. When the PDOP (*Position Dilution Of Precision*) for the visible Glonass satellites ranged between 2.5 and 2.8, the following static positioning errors were observed:

$$\sigma_x = 24.3 \text{ feet}$$
$$\sigma_y = 17.7 \text{ feet}$$
$$\sigma_z = 26.2 \text{ feet}$$

Later, when the PDOP value was in the 3.8 to 8.0 range, the three mutually orthogonal positioning errors had the following values:

$$\sigma_x = 34.4 \text{ feet}$$
$$\sigma_y = 17.4 \text{ feet}$$
$$\sigma_z = 62.3 \text{ feet}$$

A typical Glonass receiver is about four times bigger than a modern full-size 5-channel GPS receiver. With miniaturization of the electronics components using modern LSI circuit chips, its size and weight will undoubtedly shrink.

The FAA's Joint Research Efforts with Soviet Scientists

In October 1990 officials from the FAA (Federal Aviation Administration) traveled to Moscow to meet with their counterparts at the Soviet Ministry of Aviation. The purpose of this highly publicized and stimulating junket was to coordinate the exchange of technical information and attempt to build practical and efficient dual-capability receivers that can be adopted by a worldwide user community.

The Airlines Electronics Committee also instituted GPS/Glonass data interchanges to develop the proper form, fit and function for tomorrow's dual-capability receivers. In July 1990 a meeting was convened at Minneapolis between Russian scientists and members of the Airline Electronics Committee, who worked on a different design for a combined receiver.

Other Attempts to Build Dual-capability Receivers

Researchers at the Magnavox Corporation in Torrance, California, have reportedly delivered a prototype dual-capability receiver to M.I.T. with a demonstrated accuracy of 100 to 130 feet. Northwest Airlines and Honeywell Corporation are also working together on a joint receiver design. They have signed a Memorandum of Understanding with representatives of the Commonwealth of Independent States for the cooperative development of a combined GPS/Glonass receiver. Honeywell thus became the first U.S. company with the official right to market combined receivers for commercial aviation users throughout the Western world.

ARINK, Inc. is charged with the responsibility for defining the form, fit, and function for the combined receiver. The Soviets have agreed to help conduct some of the research and to promote the new receiver among their comrades in the user community within Eastern Europe. The purpose of this large-scale cooperative effort is to produce and test a successful dual-capability receiver that company engineers at Northwest Airlines will test aboard one of their jet-powered planes.

Integrity Monitoring Techniques

Like their American counterparts, Russian scientists are keenly interested in developing simple and practical integrity monitoring techniques for their spaceborne navigation system. They are convinced that present integrity monitoring methods using software routines aboard their satellites can make important contributions toward effective integrity assurance. This optimistic stance seems to persist despite the fact that American engineers have largely abandoned any hopes for reliable onboard self-monitoring techniques because of the cost and complexity of implementing it on a useful scale.

Alternate techniques for integrity monitoring (see Chapter 12) include monitoring the signals with ground-based monitor stations and/or monitoring the signals onboard the airplanes and other vehicles performing the navigation solutions. This could, in theory, be accomplished by comparing the navigation solutions obtained using different combinations of the satellites in various batches of four.

Interoperability with Other Radionavigation Systems

Interoperability between the GPS and the Glonass can bring important operational benefits, but the GPS can also be rigged for interoperability with ground-based radionavigation systems, such as Loran C and Omega, or with other space-based systems, such as Transit and the European NavSat (if it is ever built). A proper marriage between the GPS and Loran C could, for instance, provide substantial improvements in coverage while the Navstar constellation is being launched into space.

During that interval, the GPS can provide global but not continuous coverage, and the Loran C can provide continuous coverage, but not global. Navigation solutions combining the signals from the Loran C's ground-based transmitters with the spaceborne transmitters in the GPS constellation could thus extend the coverage *area* of Loran C and coverage *time* of the GPS. Of course, as more and more GPS satellites are successfully launched into space, this cooperative strategy will become increasingly academic. Even then, however, the Loran system might be used for integrity monitoring of the GPS signals or as a backup for the GPS constellation in case of large-scale system failures.

Eastport International's Integrated System for Underwater Navigation

Staff engineers at Eastport International have perfected an interesting method for achieving interoperability between the GPS satellites and a completely different kind of navigation system. Their cleverly conceived architecture uses the GPS signals to help supply accurate positioning information on remotely piloted submersible vehicles tooling around under the seas. The submersible craft employs sonar waves for underwater navigation, while the surface vessel tending it from above uses the GPS to fix its own position. Thus, both a spaceborne navigation system and a suboceanic sonar system cooperate to track the movements of the submersible craft.

The heavy seafloor beacons used in positioning the remotely piloted submersible are dropped over the side of the surface vessel. Once they sink to the bottom, they act like little underwater lighthouses to fix its position. A CRT screen in the surface ship pinpoints the location of the ship and the location of the submersible as it moves along through the water.

In 1988 Eastport International's dual-capability positioning system was pressed into action when the John Cabot search vessel was assigned to scour the ocean depths looking for Air India's crash debris off the coast of Ireland. As Figure 9.3 indicates, the debris at the 747 crash site was scattered over a 30 square-mile area in water depths of 6,000 feet. During the search, 40-knot winds accompanied by 15-foot waves lashed out against the John Cabot, whipping back and forth on the surface of the angry sea.

A typical search of this type begins when the seafloor sonar transponders are lowered over the side of the search vessel. They quickly sink to the bottom, where they are automatically anchored to the seafloor by short tethers. Once they are in place, they are rigged with positive buoyancy, which causes them to float upward, so their signals will not be blocked by seafloor obstructions.

Fragments of Air India's 747 ranged from its entire tail down to baseball-sized objects scattered across an irregular 30–square mile area at the bottom of the North Sea. Eastport International's integrated navigation system can

EASTPORT INTERNATIONAL'S INTEROPERABLE SYSTEM FOR UNDERWATER NAVIGATION

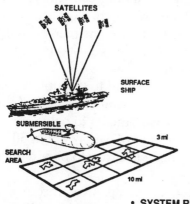

- **THE DEBRIS RANGED OVER A 30 MI^2 AREA**
    - **WATER DEPTH = 6000 FEET**
    - **WINDS = UP TO 40 KNOTS**
    - **WAVE HEIGHT = UP TO 15 FEET**
- **FRAGMENTS RANGED FROM A VIRTUALLY INTACT AIRCRAFT TAIL TO BASEBALL-SIZE OBJECTS**

- **SYSTEM RESOLVES, RECORDS, PLOTS, AND DISPLAYS:**
    - **THE POSITION OF THE REMOTELY OPERATED VEHICLE AND SURFACE SHIP**
    - **THE POSITIONS OF ANY OBJECTS LOCATED SO FAR**
    - **THE POSITIONS OF THE UNDERWATER NAVIGATION TRANSPONDERS**

Figure 9.3 Eastport International has perfected an underwater positioning system that combines the capabilities of the Navstar GPS with an underwater sonar-based system that uses seafloor transponders. In this particular application, the John Cabot search vessel is attempting to find and mark the crash debris from an Air India 747 that plunged into the North Sea off the coast of Ireland in waters 6,000 feet deep.

be rigged to resolve, record, and display the current location of the remotely piloted vehicle on the video screen, together with the location of the surface ship. The video monitor also displays the position of any debris objects located so far, any operator-designated waypoints and track lines, and the positions of the various seafloor transponders.

Two methods have been perfected to pinpoint the locations of the seafloor transponders for coordinate-system calibration and initialization. In the first method, the ship steams around the transponder field, pausing periodically to send out interrogation pulses to collect six or more slant ranges from various positions on the surface of the sea. The slant range to a particular transponder equals half the two-way signal travel time multiplied by the speed at which sound travels through seawater. Throughout the interrogation interval, the surface ship keeps track of its own position by picking up the L-band signals from the GPS satellites.

A localized coordinate system is constructed as follows: The location of transponder number 1 is regarded as a benchmark located at the origin of the coordinate system, so its horizontal coordinates x and y are both set equal to 0. Its depth, however, is regarded as an unknown. For this reason, at least one slant range measurement is necessary to establish the depth of transponder number 1. Transponder number 2 is defined as lying along the y axis of the coordinate system, so its x coordinate is equal to 0. Only its y coordinate and its depth are unknown. At least two additional slant range measurements are required to establish these two missing coordinates. All three coordinates for transponder number 3 are unknown, so at least three more slant ranges are required to nail down its three-dimensional position. Thus, at least six slant range measurements are needed to establish the relative locations of all three seafloor transponders.

Of course, random noise and timing errors tend to corrupt the measurements. Consequently, in most cases, an overdetermined system of equations is set up, with large numbers of extra measurements to help minimize the effects of statistical variations. For a typical application, about 50 different slant range measurements are usually made in order to obtain the desired level of redundancy in the solution.

To help simplify the time-consuming, labor-intensive process of seafloor transponder calibration, a second method using so-called "intelligent" transponders has been perfected. Again, the transponders are dropped off the edge of the ship. But once they have anchored themselves to the floor of the ocean, the surface ship sends down a digital pulse stream that commands each transponder to interrogate each of its neighbors. This allows the system to measure all the relevant transponder-to-transponder slant range distances. These various slant range measurements are then automatically telemetered to the surface ship, where they are processed by onboard computers. This automated approach reduces the calibration errors while saving substantial amounts of manpower and money.

When the remotely piloted submersible is ready to be tracked, the surface ship triggers another transponder attached to it, which, in turn, interrogates the other seafloor transponders. An overdetermined least squares solution provides its current best estimate of position.

The engineers and technicians at Eastport International have been highly successful in using seafloor transponders for tracking remotely piloted submersibles. Their system, which is surprisingly versatile and accurate, features clever design techniques and user-friendly displays.

10

The Navstar Satellites

A Navstar satellite is a highly complicated but surprisingly reliable machine. Although it is constructed from 65,000 separate parts, it is designed to last 7½ years or 580 million miles, whichever comes first. Throughout its on-orbit life, it points part of itself toward the earth, part of itself toward the sun—all within half a degree accuracy.

The 12 spiral-wound helical antennas on the lower edge of the spacecraft (see Figure 10.1) blanket the full disk of the earth with a fan of nearly constant-intensity electromagnetic energy. Four helical antennas are grouped in the center quad; eight others are rimmed around the center quad in a circular pattern. Working together in partnership, all 12 antennas generate navigation signals spanning a 28-degree cone that just matches the angular dimensions of the earth as seen from the satellite's high-altitude vantage point in space.

Many of the electronic components inside the spacecraft operate most efficiently at, or near, room temperature. In part, this stems from the fact that their solid-state cousins and nephews installed in microwave ovens, digital wrist watches, commercial jetliners, and personal computers were designed primarily for terrestrial applications. The cesium and rubidium atomic clocks carried onboard the satellites are especially sensitive to unwanted temperature excursions.

Seven thermostatically actuated louvers on the two opposite sides of the spacecraft help maintain a nearly constant internal spacecraft temperature. They open and close automatically to hold in heat or allow excess heat to escape. Thermal blankets and electrical resistance heaters provide additional thermal control.

The two bat-wing solar arrays protruding from the sides of the spacecraft swivel and tilt to catch the perpendicular rays of the sun. Their sun-oriented surfaces are populated with silicon solar cells capable of generating 700

A TYPICAL BLOCK II NAVSTAR NAVIGATION SATELLITE

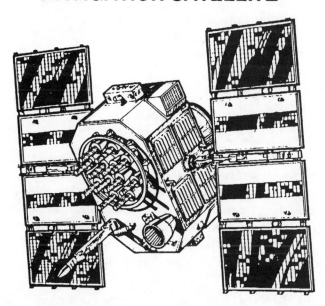

Figure 10.1   Each Navstar satellite is constructed from 65,000 separate parts, yet it is designed to operate 7 ½ years or 580 million miles—without servicing! The solar cells covering the bat-wing solar arrays generate 700 watts end-of-life electrical power to drive the onboard transmitters and the other vehicle subsystems. Internal temperatures are maintained, within narrow tolerances, by thermostatically controlled louvers and insulation blankets wrapped around some of the more critical spacecraft components.

watts end-of-life electrical power. That power is used to drive the satellite's navigation transmitters, its atomic clocks, its momentum wheels, and various other onboard subassemblies. When a Navstar satellite first arrives in space, its solar cells generate higher power levels. But, after it lumbers around in orbit for 7 ½ years, in a rather intense portion of the upper Van Allen Radiation Belt, its power-generating capacity is degraded by 15 or 20 percent.

Surveyors and pilots, scientists, and mariners clutching their sleek Navstar receivers seldom ever ponder the exquisite design and painstaking craftsmanship that are built into each satellite arching across the sky. But, without the fierce dedication of a hundred dozen trajectory experts, test engineers, component inspectors, and launch crews, too, the Navstar revolution could never have presented so many fantastic benefits to the people living on planet earth.

The Eight Major Spacecraft Subsystems

Aerospace engineers simplify their complicated design studies by breaking their satellites into separate but interacting subsystems, each of which performs a clearly defined function. Eight major navigation-related subsystems are installed onboard each Navstar satellite:

1. Orbit injection subsystem
2. Tracking, telemetry, and command
3. Attitude and velocity control
4. Electrical power
5. Navigation subsystem
6. Reaction control
7. Thermal control
8. Structures and mechanisms

A few other spacecraft subsystems and functional components are also carried onboard, but these eight subsystems are the ones that perform all of the functions directly related to the satellite's fundamental mission of providing accurate navigation to thousands of enthusiastic users on the ground below. In the paragraphs to follow, these eight major spacecraft subsystems will be introduced and discussed one by one.

The Orbit Injection Subsystem

The orbit injection subsystem consists of two solid rocket motors that hurl the GPS satellite from its low-altitude parking orbit up to its final destination orbit 10,898 nautical miles above the earth. Those two rockets are called the *perigee-kick motor* and the *apogee-kick motor* because they are ignited to add extra velocity at both the perigee and the apogee of the elongated transfer ellipse.

The Pam-D *perigee-kick motor* is ignited shortly after booster burnout. The energy it provides drives the spacecraft onto its transfer ellipse, so it can coast up to apogee 180 degrees away. When it arrives at apogee, the apogee-kick motor, a Star 37-XF solid rocket built by Thiokol, is ignited to achieve circularization. When the Pam D burns out, its casing is discarded, but the empty casing of a Star 37-XF remains permanently attached to each Block II Navstar satellite.

Large solid rockets are not as efficient as their liquid rocket counterparts, but they are simpler and more compact and can be stored in orbit with relative safety over long periods of time.

Tracking, Telemetry, and Command

The tracking, telemetry, and command subsystem includes the black boxes and the S-band antennas that allow the ground control segment to communicate with the spacecraft while it is orbiting the earth. Fresh ephemeris constants and clock correction factors are transmitted to each satellite to-

Booster Rocket Pioneers

The rockets that hurl the Navstar satellites into orbit are direct descendents of the highly destructive Chinese "fire arrows" built and launched by Chinese military engineers 750 years ago. The earliest Chinese rockets were slender tubes stuffed with gunpowder and fastened to long flat sticks that jutted out behind the rocket to promote stable flight. In 1232 they were launched in large quantities on the outskirts of Peking, when special Chinese rocket brigades successfully pushed back Mongol cavalrymen. And, in 1249, they were used with great effect by the Moors in their military campaign along the Iberian Peninsula.

Near the beginning of the nineteenth century, Englishman William Congreve concocted superior powder blends and moved the stabilizing stick to the center of the rocket for improved accuracy. In 1807 the British blasted Copenhagen with 25,000 Congreve rockets. Seven years later, when they bombarded Fort McHenry, they inadvertently provided "the rocket's red glare," which helped inspire America's National Anthem.

In 1903 a lonely Russian schoolteacher, Konstantin Tsiolkovsky, correctly concluded that rockets fueled with liquid hydrogen and liquid oxygen would be considerably more efficient than the simpler solid-fueled rockets then in use. He also devised a concept for stacking rockets one atop the other to yield the enormous speeds necessary for successful interplanetary travel.

Twenty-three years later Dr. Robert Goddard knelt on the frozen ground in his Aunt Effie's cabbage patch at Auburn, Massachusetts, and casually used a blowtorch to ignite the world's first liquid-fueled rocket. Goddard is today revered for his expansive expertise, but during his lifetime his contemporaries criticized him unmercifully because he had once dared to mention the possibility of sending a small flash powder to impact the moon. Years later, when one of his liquid-fueled rockets reached its design altitude of 2,000 feet, a banner headline wryly commented: "Moon Rocket Misses Target by 237,799½ Miles!"

The rockets built by the Goddard team were all handcrafted machines, but Germany's rocketeers, working under the direction of Werner von Braun, constructed liquid-fueled rockets in mass-production quantities. When World War II fizzled to a halt, many of the German scientists came to Los Alamos to help American's military and space-age rocketeers.

In 1961, when President Kennedy courageously announced that the United States would conquer the moon, America's rocketeers had not yet orbited a single astronaut. The Saturn V moon rocket they later developed for the mission, was the pinnacle of the rocket maker's art. But it was expendable; NASA's space shuttle is a *reusable* booster. It delivers payloads weighing as much as 50,000 pounds and brings others back to earth for refurbishment and repair, gently landing—as TV Newsman Edwin Newman once observed—"like a butterfly with sore feet."

---

gether with health status messages. This information is packed into the 50-bit-per-second data stream for broadcast back down to the users on the ground.

Two different S-band antennas are carried onboard each Navstar satellite:

1. A forward conical spiral mounted atop a bicone horn
2. An aft conical spiral

These two spear-like antennas point in opposite directions forward and aft, because, as the spacecraft coasts around the earth, it twists and turns. Thus, two oppositely oriented S-band antennas are necessary to maintain strong, distortion-free communications with the 16-foot upload antennas positioned at widely scattered positions on the ground below.

The uplink transmissions for the tracking, telemetry, and command subsystem are centered at 2,227.50 megahertz in the S-band portion of the frequency spectrum. The downlink transmissions are centered at 1,783.74 megahertz. The downlinks are used primarily for message acknowledgements and parity-checking responses to let the ground crews know that each uplink message has been successfully received.

Attitude and Velocity Control

The attitude and velocity control subsystem (see Figure 10.2) helps maintain a continuous earth-seeking vehicle orientation to allow the spacecraft's helical antennas to blanket the full disk of the earth with constant-intensity L-band navigation signals. The spacecraft attitude is maintained by selectively adjusting the rotation rates of the spinning momentum wheels, which react against the spacecraft to create precise controlling torques. The four

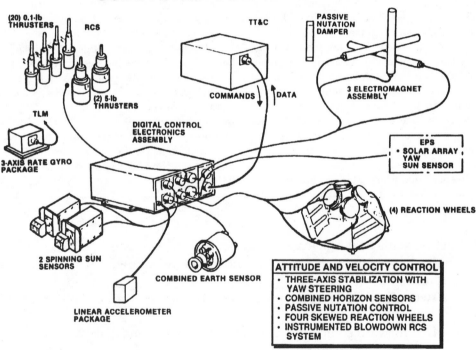

Figure 10.2 The attitude and velocity control subsystem helps a GPS satellite maintain its desired earth-seeking orientation in space. This is accomplished by systematically adjusting the rotation rates of the four momentum wheels arranged in "teepee" fashion. During these maneuvers, the rotation rate of each momentum wheel gradually increases. Consequently, its momentum must be "dumped" by selectively activating the three mutually orthogonal electromagnets as they cut across the earth's magnetic field.

momentum wheels are arranged in "teepee" fashion. This arrangement provides multiple redundancy to cover for any momentum wheel failures that may occur. If any one of the four wheels fails, the other three can still maintain the desired vehicle orientation.

As the momentum wheels make repeated adjustments in vehicle attitude, they begin to spin faster and faster. If their momentum is not periodically reduced (dumped), the wheels will begin to shimmy and vibrate. Later, if nothing intervenes, they will fly apart due to increasing centrifugal force. The spinning momentum wheels can be slowed by firing the attitude control thrusters in static couples to create carefully controlled reaction forces. This approach works as intended, but it burns up the satellite's precious hydrazene propellants.

Another way to dump momentum is to selectively activate the three mutually orthogonal electromagnets in the upper right-hand corner of Figure 10.2. When one of the electromagnets is activated, it cuts across the earth's magnetic lines of flux much like an electric motor or a dynamo. The resulting torque on the spacecraft reacts against the spinning wheels to dump their momentum. This clever engineering approach uses the electricity generated by the satellite's solar cells, but it does not consume any valuable hydrazene propellants.

The earth sensor and the two spinning sun sensors provide continuous measurements to help pinpoint the precise locations of earth and sun. These measurements are routed into the appropriate feedback control loops to trigger subtle and continuous adjustments to help maintain the desired orientations for the main body of the spacecraft and its two solar arrays.

Electrical Power

The electrical power subsystem includes two large, flat solar arrays, each covered on a single side with silicon solar cells. [1] The entire spacecraft swivels, and the solar arrays tilt to intercept the perpendicular rays of the sun for photovoltaic conversion into electrical power. The electricity generated by the vehicle's solar cells is used to power the other onboard subsystems, including its momentum wheels, its L-band and S-band transmitters, its cesium and rubidium atomic clocks, and its electrical resistance heaters.

Three 15-ampere-hour nickel-cadmium batteries are used to store excess electrical power. This extra power covers for peak load demands and allows the spacecraft to transmit uninterrupted navigation signals when it is periodically shrouded in darkness within the earth's shadow. During a year in orbit, each satellite experiences shadowing periods during two separate

---

[1]Small gaps unpopulated with solar cells were purposely left on the solar arrays for the Block II satellites just in case some later version of the spacecraft might require large amounts of extra power. Up to 1,000 watts end-of-life electrical power can thus be provided without major hardware modifications.

"eclipse seasons," roughly six months apart. Temporary eclipses occur whenever the edge of the orbit plane points in the general direction of the sun. Averaged over the entire year, each GPS satellite is in the earth's shadow less than 1 percent of the time.

Navigation Subsystem

The navigation subsystem includes four quadruple redundant atomic frequency standards: two cesium atomic clocks, and two rubidium atomic clocks. These onboard clocks generate the evenly spaced C/A- and P-code timing pulses that are broadcast down to the users on the ground.

An active base-plate thermal control unit mounted on the base of each rubidium atomic clock helps maintain a nearly constant internal temperature. This helps the clocks keep time with enhanced stability. The onboard atomic clocks are never operated simultaneously if one fails, another is switched on to take its place.

Reaction Control

The reaction control subsystem is powered by storable hydrazene propellants. At insertion, two 5-pound trim thrustors are used in making minor adjustments to the spacecraft orbit. In addition, twenty 0.1-pound attitude control thrustors act as a backup for momentum dumping in case electromagnet failures occur during the mission. The thrustors, fired in the proper combinations, to obtain a small translational ΔV are also used to maintain nearly equal satellite spacing within the six orbital rings.

Hydrazene is not very efficient for providing thrust, but it is a simple monopropellant that is storable and largely non-contaminating. When it is chemically decomposed to produce thrust, its principal by-products are water and ammonia. When burned in its steady-state mode, the specific impulse of hydrazene is approximately 220 seconds. The specific impulse of a rocket propellant combination can be defined as *the number of seconds during which a pound of the propellant will produce a pound of thrust*. The higher the specific impulse, the more efficient the propellant. Some practical liquid bi-propellants (fuel and oxidizer stored in separate tanks) are nearly twice as efficient as hydrazene, but they entail various undesirable properties, such as corrosive by-products and cryogenic storage requirements.

Thermal Control

The satellite's thermal control subsystem employs a variety of highly developed design techniques to help maintain a nearly constant internal spacecraft temperature. For most space vehicles an internal "shiftsleeve" environment close to the human comfort zone is highly desirable. This is true because many of the internal spacecraft components (or their predecessors) were originally designed to operate in room temperature environments on or near the earth.

Seven thermostatically controlled thermal louvers resembling venetian blinds in appearance and function are mounted on the two opposite sides of the spacecraft. They are rigged so that they open and close automatically to let heat out or keep heat in. The thermal control louvers are actuated by spiral-wound bi-metallic strips composed of two dissimilar metals that expand and contract differentially when subjected to thermal loads. These differential expansions and contractions cause the bi-metallic strip to straighten out or coil tighter, thus opening or closing the thermal louvers. An ordinary household thermostat operates in a similar manner. The seven thermal louvers are all independently actuated; if one fails, the other six are still capable of maintaining the desired internal spacecraft temperature.

Goldized mylar-kapton insulation blankets help provide additional thermal control. Mylar and kapton are space-age plastics similar to the transparent garment bags used in drycleaning establishments. Some of the multi-layer goldized insulation wrappings on the GPS spacecraft are 13 layers thick. According to the thermal experts who masterminded the spacecraft design, if they carefully wrapped a 1-pound block of ice in a 13-layer mylar-kapton insulation blanket and placed it on the end of the Santa Monica pier, one year later they would still find ice inside!

Thermostatically controlled resistance heaters positioned at certain specific locations inside the spacecraft also help maintain a stable internal temperature. The active base-plate thermal control unit that selectively heats the base of the rubidium atomic clocks is one such resistance heater.

Structures and Mechanisms

The structure that houses and support the delicate spacecraft components consists primarily of aluminum-bonded honeycomb panels with thin aluminum sheets attached to hollow hexagonal cells copied from the vastly underrated bumblebee.[2] The six-sided aluminum honeycomb cells are bounded top and bottom with thin, flat aluminum panels to provide a lightweight but extremely rigid structure. An exploded view of the structural components that hold the satellite together is presented in Figure 10.3.

---

[2]The two British biologists Kark Vogt and Ernst Haeckel once observed that the British Empire owed its power and wealth largely to bumblebees. They reasoned that the true source of England's influence resided mainly in its superb Navy, whose sailors were fed with beef "which came from cattle that subsisted on clover, which would not grow without pollination by bumblebees." Vogt and Haeckel did not live long enough to learn how strongly today's space program has come to depend on the technology pioneered by the industrious little bumblebee.

STRUCTURAL SUBSYSTEM EXPLODED VIEW

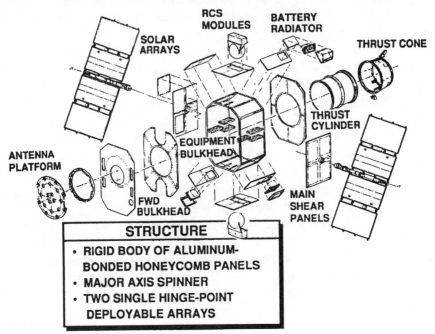

STRUCTURE

- **RIGID BODY OF ALUMINUM-BONDED HONEYCOMB PANELS**
- **MAJOR AXIS SPINNER**
- **TWO SINGLE HINGE-POINT DEPLOYABLE ARRAYS**

Figure 10.3 Aluminum-bonded honeycomb panels, copied from the lowly bumblebee, make up the main structural elements of the Navstar satellite, with two big solar arrays protruding from its sides. As the spacecraft travels toward its final destination, the solar panels are folded against its sides. When it arrives, spring tension is released to deploy the arrays into their desired wing-like configuration.

The only single-point failure mechanism on the GPS satellite is the device used in deploying its winglike solar arrays. If that mechanism fails, there are no backups; the mission is lost. In a 1-g environment the deployment mechanism is scarcely strong enough to support the weight of the solar array. But, in the weightlessness of outer space, its strength is definitely sufficient. The two single hinge–point deployment mechanisms are maintained under constant spring tension. When the spring tension is released, the solar arrays deploy automatically. So far, all of the solar array deployment mechanisms have worked successfully without any glitches in space.

Each GPS Block II satellite weighs approximately 2,000 pounds. Its solar arrays cover 78 square feet of surface area to produce ample quantities of electrical power. The design goal for the Block II mission calls for a 7.5-year on-orbit life, but the spacecraft must carry sufficient consumables for a 10-year mission. Throughout the Navstar Program, the spacecraft subsystems and components have performed exceptionally well. Some Block I satellites—which were designed for a 4-year mean-mission duration—have lived 10 years and more in the punishing environment of outer space.

On-orbit Test Results

The atomic clocks onboard the GPS satellites have turned out to be surprisingly stable and accurate. Their fractional frequency stability was originally set at one part in a trillion or 10^{-12}. The fractional frequency stability, a dimensionless quantity, equals the 1σ error pulse-to-pulse divided by the duration between pulses. A clock with a fractional frequency stability equal to 10^{-12} would theoretically lose or gain only one second every 30,000 years. For the early Block I satellites this seemingly stringent performance specification was consistently exceeded by a factor of 2 to 5. So, for the later missions (starting with Navstar 9), a more stringent target value of 2×10^{-13} was established by DoD mission planners.

To help insure that this much more demanding goal would be successfully achieved, the spacecraft designers introduced their active base-plate thermal control unit to help minimize the slight internal temperature variations being experienced by the atomic clocks. The active base-plate thermal control unit heats the base of the atomic clocks whenever their temperature drops below a preset level. The average temperature of the clocks thus increases by a few degrees, but their temperature variations turn out to be about 10 times smaller.

Excellent performance for the reaction control subsystem has also been achieved. All of the GPS satellites have reached their destination orbits with ample quantities of hydrazene propellants. And once they are in orbit, they have been consuming the remaining propellant very sparingly. In part, this stems from the fact that all of the recent momentum dumps have been accomplished by selectively activating the three mutually orthogonal electromagnets, rather than by using the vehicle's attitude control thrustors.

The three nickel-cadmium batteries on each satellite have consistently operated within performance specifications. The spacecraft batteries have seldom been discharged as much as 35 percent.

The attitude, velocity and control subsystem has also performed with noteworthy efficiency. Military performance specifications call for earth-seeking orientation errors not to exceed half a degree. The actual attitude errors have usually been 2 to 5 times smaller than these rather stringent specifications. However, some momentum wheels aboard the Block II satellites—on the ground and in space—have experienced difficulties with "lubricant starvation." This unexpected anomaly apparently stems from the fact that, when the Shuttle Challenger disaster occurred, the GPS satellites awaiting launch had to remain in storage for extended intervals. This seems to have caused evaporation of the lubricants and/or changes in their chemical composition. At this writing, various solutions are being explored.

The Multiyear Spacecraft Procurement

The GPS Block II satellites were purchased at surprisingly affordable rates because the Department of Defense was willing to sign a $1.17 billion multiyear procurement contract to purchase 28 of them over a period of several years. This large guaranteed procurement reduced the cost of each satellite by an estimated 15 percent, compared with the DoD's more conventional year-by-year spacecraft commitments.

Some of the savings stemmed from the fact that the large block buy allowed parts and materials to be ordered in much larger quantities, thus promoting the economies of scale. Significant shipping and handling efficiencies could also be realized, and set-up costs for the various parts could be spread over many more units. Raw materials, such as aluminum sheets and bars, were also a bit cheaper when purchased in bulk quantities.

Another substantial saving came about because the prime contractor, Rockwell International, could afford to install automated testing equipment to help reduce the estimated 25-percent testing expenditure associated with a typical spacecraft procurement. An $8 million thermal vacuum chamber, for instance, was installed at the Rockwell facility in Seal Beach, California, to help foster efficient testing of the production satellites under realistic vacuum conditions.

The thermal vacuum chamber is a large metal cylinder 37 feet long and 27 feet in diameter. After several hours of pumping, the atmospheric pressure inside is equivalent to 1.3×10^{-10} atmospheres. This corresponds to the ambient pressure that a satellite experiences 150 nautical miles above the earth. The GPS satellites go much higher, where the air is even thinner, but 1.3×10^{-10} atmospheres is quite sufficient for the test procedures required.

A large anechoic (anti-echo) chamber is also used in testing the Navstar satellites. Attempting to duplicate the transmissions coming to and from a GPS satellite in an ordinary room is totally impractical because any signals reflecting from the walls and ceiling tend to invalidate the accuracies of the test. The anechoic chamber eliminates most stray reflections because its inner surfaces are covered with big plastic spikes that protrude into the test chamber and absorb any incident L-band or S-band electromagnetic waves. This helps the test engineers accurately reproduce the benign environment for the GPS signals as they will ultimately be transmitted and received in outer space.

A second multiyear procurement contract for the Block IIR (replenishment) satellites has also been signed by the DoD. General Electric captured this contract, which authorized them to build 20 replenishment satellites for launch into orbit in the mid-1990s and beyond. Substantial cost savings also resulted from this large multiyear spacecraft procurement.

The Rocket's Red Glare

Both rockets and jets are based on the same principle that causes a toy balloon, carelessly released, to loop around in tight kamikaze spirals in your living room. A jet engine obtains its oxygen from the surrounding air, but a rocket carries its own oxidizer onboard so that it can operate in the vacuum of space. The oxidizer can be contained in a separate tank, mixed with the fuel, or embedded in oxygen-rich compounds.

A large *liquid* rocket is usually rigged with two separate tanks—one for the fuel and the other for the oxidizer. The two fluids are pumped or pushed under pressure into a small chamber above the exhaust nozzle, where burning takes place.

A *solid* rocket is like a slender tube filled with gunpowder; the fuel and oxidizer are mixed together in a rubbery cylindrical slug called the grain. Solid propellants are not pumped into a separate combustion chamber; burning takes place along the entire length of the cylinder. For this reason, the tank walls must be built strong enough to withstand the combustion pressure.

Rocket design decisions are dominated by the desire to produce the maximum possible velocity when the propellants are burned. A booster rocket's velocity can be increased in three principal ways: by using propellants with high efficiency, by making the rocket casing and its engines as light as possible, and by stacking increasingly smaller stages one atop the other.

More efficient propellants unfortunately tend to exhibit highly undesirable physical and chemical properties. Hydrogen is a good fuel, but it can spark devastating explosions. Liquid oxygen is a good oxidizer, but it will freeze all lubricants and crack most seals. Fluorine is even more efficient, but it is so reactive that water will not quench it and it can actually cause metals to burn.

Miniaturized components, special fabrication techniques, and high-strength alloys can all be used to shave excess weight. But as one rocket pioneer wryly observed, "a designer can add only so much lightness." The solution is to employ a "staging" technique in which a series of progressively smaller rockets are stacked one atop the other. A multistage rocket cuts down its own weight as it flies along by discarding large tankage and heavy engines. However, orbiting even a small payload with a multistage rocket requires a huge booster. The Saturn V moon rocket, for example, outweighed its Apollo payload by a factor of 60 to 1.

Booster Rockets

Starting in 1978, ten 960-pound Block I satellites were successfully carried aloft aboard Atlas F boosters launched from Vandenberg Air Force Base.[3] Before each Atlas F fell into the Pacific, its payload was separated and a Pam D perigee-kick motor was ignited to hurl the spacecraft onto a greatly elongated transfer ellipse. The satellite then coasted up to its 11,000–nautical mile apogee, where an apogee-kick motor firing circularized its orbit. The Atlas F booster was not selected to launch the 2,000-pound Block II satellites for two different reasons: (1) the Atlas F was not powerful enough and (2) there were not enough Atlas F's left to handle the upcoming missions.

The Block II GPS satellites were initially scheduled to ride into space in the Shuttle cargo bay, but, when the Shuttle Challenger disaster occurred, expendable boosters were quickly substituted for future GPS missions.

---

[2]An eleventh Atlas F booster carrying Navstar 7 exploded shortly after liftoff, thus littering the landscape at the Vandenberg Launch Complex with thousands of GPS satellite fragments.

McDonnell Douglas won the contract for the production of 20 uprated Delta II's (Medium Launch Vehicles), each capable of boosting a single Block II satellite into space. So far, the Delta II Medium Launch Vehicle has performed without failure in this newly assigned mission.

Orbital Perturbations

As the GPS satellites hurtle around the earth, subtle forces of perturbation tend to distort their simple elliptical orbits, so their ephemeris constants must be more numerous than would otherwise be required. The major perturbations acting on a GPS satellite are summarized in Table 10.1.

Notice that the biggest force of perturbation is caused by the second zonal harmonic, the earth's equatorial bulge. If the effects of this perturbation could not be accurately modeled, a 1,000-foot positioning error would be created in the satellite's orbit within one hour after the last ephemeris coordinate update was received. Fortunately, the extra ephemeris constants sent up to the satellites from the ground accurately compensate for this rather large potential source of error.

Lunar and solar gravitational effects are the second and third largest perturbations acting to distort the orbits of the GPS satellites. Left unmodeled they would create satellite positioning errors of 130 feet and 65 feet respectively, within an hour after the last update. The fourth zonal harmonic in the earth's gravity model constitutes the next largest perturbation, fol-

Table 10.1 Summary of the perturbing forces acting on a GPS satellite

Perturbation	Maximum perturbing acceleration (g's)	Dominant perturbation period (hours)	Maximum excursion growth in one hour (feet)
Earth Attraction	5×10^{-2}	12	—
2nd zonal harmonic	5×10^{-6}	secular +6	1,000
Lunar gravity	5×10^{-7}	secular +12	130
Solar gravity	3×10^{-7}	secular +12	65
4th zonal harmonic	10^{-8}	3	2
Solar radiation pressure	10^{-8}	secular +3	2
Gravity anomalies	10^{-9}	various	0.2
All other perturbations	10^{-9}	various	0.2

lowed by the solar radiation pressure, which is created whenever sunlight illuminates any exposed surface. The earth is 93,000,000 miles from the sun and, at that distance, the solar illumination exerts a pressure of only about 5 pounds per square mile. This subtle pressure accelerates each GPS satellite at only 10^{-8} g's but, despite its tiny magnitude, the solar radiation pressure creates the biggest *uncertainty* in the modeling of a GPS satellite orbit. The modeling uncertainty associated with solar radiation pressure stems from the fact that, as each satellite swings around its orbit, its main body presents a slightly different cross-sectional area to the perpendicular rays of the sun. Consequently, the forces of perturbation acting on the satellite vary in a manner that is extremely difficult to model to a high degree of precision.

The biggest orbital distortion created by the earth's equatorial bulge on a satellite's orbit is sketched in Figure 10.4. As the sketch on the left-hand side indicates, when the satellite is in the northern hemisphere, it is pulled *down* towards the equator by the earth's equatorial bulge when it is in the southern hemisphere, it is pulled *up* toward the equator. The net result is a gradual twisting of the orbit plane that causes a systematic variation in its equatorial crossing point (ascending node).

NODAL REGRESSION DUE TO THE EARTH'S EQUATORIAL BULGE

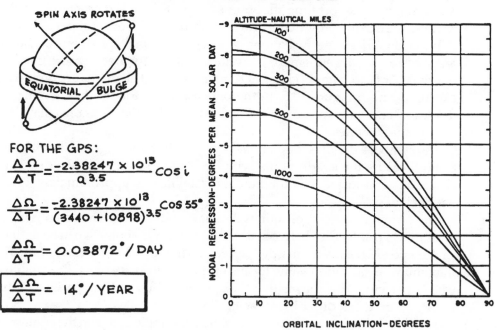

FOR THE GPS:

$$\frac{\Delta \Omega}{\Delta T} = \frac{-2.38247 \times 10^{13}}{a^{3.5}} \cos i$$

$$\frac{\Delta \Omega}{\Delta T} = \frac{-2.38247 \times 10^{13}}{(3440 + 10898)^{3.5}} \cos 55°$$

$$\frac{\Delta \Omega}{\Delta T} = 0.03872°/\text{DAY}$$

$$\boxed{\frac{\Delta \Omega}{\Delta T} = 14°/\text{YEAR}}$$

Figure 10.4 The earth's equatorial bulge (here, for clarity, depicted as a thick midriff girdle) twists a satellite's orbit so that its nodal crossing point marches relentlessly around the equator. Some low-altitude satellites experience nodal regression rates amounting to several degrees per day; the high-altitude orbits of the GPS satellites twist more slowly, on average, only about 14 degrees in a mean solar year.

The orbits of low-altitude satellites are strongly influenced by the earth's equatorial bulge. If, for instance, a satellite is launched into a circular orbit at 500 miles altitude, the equatorial bulge will cause its nodal crossing point to march around the equator 5.5 degrees per mean solar day. For satellites at higher altitudes, the earth behaves more like a "point mass" as seen from their vantage point in space, so the earth's equatorial bulge does not affect them nearly so much. The variation in the nodal crossing point for a GPS satellite totals only about 14 degrees per year. This is, incidentally, about half the rate at which the nodal crossing point of a typical GPS orbit varies due to perturbations caused by gravity from the sun and the moon.

The Spacecraft Ephemeris Constants

In the seventeenth century England's brilliant scientist and natural philosopher, Sir Isaac Newton, proved mathematically that, if a satellite is in orbit about a planet composed of concentric shells of equal-density material—and that satellite is experiencing no other forces of perturbation—six constants substituted into a few simple equations can, in theory, be used to predict its motion forevermore.

The constants usually chosen are the six classical Keplerian orbital elements: $a, e, i, \Omega, \omega,$ and T. These six orbital elements can be used to predict the motion of a planet orbiting the sun or the motion of a man-made satellite orbiting the earth—provided there are no appreciable forces of perturbation.

When significant perturbations do exist, additional ephemeris constants are necessary or the ephemeris constants must be updated at more frequent intervals, or both. The six classical Keplerian orbital elements are defined in capsule form in Figure 10.5. In the paragraphs to follow they are further discussed and their values are defined as related to the GPS constellation of man-made satellites.

The symbol a denotes the *semi-major axis* or the half-length of the elliptical orbit. Measure the length of the orbit from end to end, and then divide by two. The semi-major axis for the GPS satellite's is approximately 14,300 nautical miles.

The *orbital eccentricity*, e, defines the shape of the orbit. The eccentricity of an elliptical orbit varies between 0 (circle) and 1 (parabola). It can be defined as the apogee radius minus the perigee radius divided by their sum. The maximum specification value for the orbital eccentricity of a GPS satellite is 0.02 (2 percent). This means that the apogee radius and the perigee radius for a GPS orbit can differ by, at most, 288 nautical miles.

The *inclination*, i, of a satellite orbit equals the angle between the equatorial plane and its orbit plane. The orbital inclination for the Block II GPS satellite is 55 degrees. For the Block I satellites it is 63 degrees.

The symbol Ω denotes the *ascending node* of the satellite's orbit or the inertial location at which the satellite crosses the equator as it moves from the southern hemisphere into the northern hemisphere. The location of that

THE SIX KEPLERIAN ORBITAL ELEMENTS

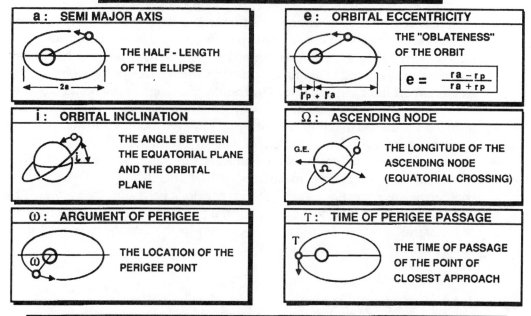

THE SIX KEPLERIAN ORBITAL ELEMENTS
a, e, i, Ω, ω, T

a : SEMI MAJOR AXIS — THE HALF - LENGTH OF THE ELLIPSE

e : ORBITAL ECCENTRICITY — THE "OBLATENESS" OF THE ORBIT

$$e = \frac{r_a - r_p}{r_a + r_p}$$

i : ORBITAL INCLINATION — THE ANGLE BETWEEN THE EQUATORIAL PLANE AND THE ORBITAL PLANE

Ω : ASCENDING NODE — THE LONGITUDE OF THE ASCENDING NODE (EQUATORIAL CROSSING)

ω : ARGUMENT OF PERIGEE — THE LOCATION OF THE PERIGEE POINT

T : TIME OF PERIGEE PASSAGE — THE TIME OF PASSAGE OF THE POINT OF CLOSEST APPROACH

- THESE SIX ORBITAL ELEMENTS ARE MATHEMATICALLY EQUIVALENT TO 3D POSITION AND VELOCITY (X,Y,Z, Ẋ,Ẏ,Ż)
- THE GPS EPHEMERIS COORDINATES INCLUDE 4 EXTRA ELEMENTS FOR CLOCK ERRORS AND ABOUT 10 OTHERS FOR ORBITAL PERTURBATIONS

Figure 10.5 In the absence of perturbations, the six classical Keplerian orbital elements characterize a satellite's orbit so precisely that they could, in theory, be used to predict its future location forevermore. In practice, of course, uncertainties and forces of perturbation do exist, so future locations can be predicted only by using larger numbers of constants, which must be updated more frequently.

nodal crossing point is measured in the equatorial plane from the Vernal Equinox to the nodal crossing point. The 24 Block II satellites have six different inertial nodes, one for each of the six orbit planes.

The *argument of perigee,* ω, is the angle between the ascending node and the perigee of the orbit measured in the orbit plane. For the GPS satellites, the argument of perigee cannot be predicted in advance because the mission specialists target for circular orbits.

The *time of perigee passage,* T, is the time at which the satellite reaches its point of closest approach to the center of the earth. Before liftoff, the value of T is unpredictable, again, because the mission specialists target for circular Navstar orbits.

In a perturbation-free environment, these six classical Keplerian orbital elements could, in theory, be used to project the position of each Navstar satellite weeks, months, or even years into the future. This is in keeping with Newton's laws of classical mechanics, because knowing these six orbital elements is tantamount to knowing the three initial position coordinates (x_0, y_0, z_0) and the three initial velocity components ($\dot{x}_0$, $\dot{y}_0$, $\dot{z}_0$) plus all the forces (e.g., the earth's gravitational attraction) acting on the satellite.

In actual practice, of course, perturbations do exist, so 16 ephemeris constants are needed to define the orbit of each GPS satellite. As Table 10.2 indicates, some of them, such as eccentricity and semi-major axis, have been previously discussed. Others include the *rate of change* of some of the Keplerian orbital elements (e.g., $\dot{\Omega}$) the rate of change of the nodal crossing point, and $\dot{i}$, the rate of change of the inclination.[3] Other perturbations are induced by the earth's higher-order potential harmonics and the gravitational fields of the sun and the moon. The six coefficients (C_{uc}, C_{us}, C_{rc}, C_{rs}, C_{ic}, and C_{is}) are used to model the effects of these gravitational perturbations. The 16 ephemeris constants listed in Table 10.2, plus 4 polynomial coefficients defining each satellite's current clock behavior, are uploaded to each satellite in packets from the ground once per day. Those same packets of constants are then broadcast back down in the 50-bit-per-second data stream for use by ground-based receivers.

Satellite Viewing Angles

As seen from the GPS altitude of 10,898 nautical miles, the earth subtends an angle of approximately 27.8 degrees. This value is valid for a 0-degree mask angle. When a larger mask angle is imposed, the users within a thin

Table 10.2 The GPS satellite ephemeris constants in the 50-bit-per-second data stream

M_0	*Mean anomaly*
n	Mean motion difference
e	Eccentricity
$\sqrt{a}$	$\sqrt{}$ of semi-major axis
Ω_0	Right ascension
i_0	Inclination
ω	Argument of perigee
$\dot{\Omega}$	Rate of right ascension
$\dot{i}$	Rate of inclination
C_{uc}, C_{us}	Correction terms to arg. of latitude
C_{rc}, C_{rs}	Correction terms to orbital radius
C_{ic}, C_{is}	Correction terms to inclination
t_0	Ephemeris reference time

---

[3]Notice that the rate of change of inclination is denoted by $\ddot{i}$ (the letter i with two dots on top!)

circular rim near the outer edge of the earth cannot use the signals from that satellite for accurate navigation. For a 10-degree mask angle, the useable portion of the earth subtends a 27.4 degree angle, as seen from the GPS satellite.

Earth-shadowing Intervals

The eclipse intervals of a satellite—and the total amount of time it spends in sunlight—have important impacts on the design of its solar arrays and its storage batteries. When the edge of its orbit points toward the sun (β angle = 0), a GPS satellite passes along a full cord of the earth as seen from the sun. This gives the maximum eclipse interval of 56 minutes in the earth's shadow.

When the β angle (angle between the radius vector to the sun and orbit plane) lies between ± 13 degrees, the spacecraft passes into a portion of the earth's shadow for various intervals ranging from 0 to 56 minutes, as indicated by the graph in the lower right-hand corner of Figure 10.6. For

EARTH SHADOW CHARACTERISTICS

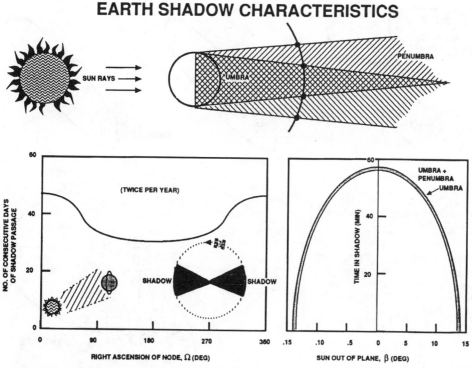

Figure 10.6 The maximum eclipse interval (56 minutes) for a GPS satellite occurs when the edge of its orbit plane points directly toward the sun. This happens at the middle of two "eclipse seasons," roughly six months apart, during which eclipses occur twice per day for 35 to 45 successive days. Averaged over the entire year, each GPS satellite spends more than 99 percent of its time fully illuminated by the sun.

each satellite, the β angle reaches 0 degrees twice per year at the center of two "eclipse seasons" about six months apart. The graph in the lower left-hand corner of Figure 10.6 defines the relationship between the location of the ascending node (equatorial crossing point of the satellite) and the duration of the eclipse season. Notice that, depending on the node (Ω) of the satellite orbit, the eclipse season lasts somewhere between 35 and 45 days.

Repeating Ground-trace Geometry

As the sketches at the top of Figure 10.7 indicate, the orbital period of a GPS satellite is 12 hours, as measured in inertial space. But, because the earth rotates under its orbit plane, the satellite traces out the same repeating ground trace approximately every 24 hours. As the satellite orbits the earth, it travels along its orbit 30 degrees per hour, but the earth rotates under it at 15 degrees per hour (one 360-degree revolution every 24 hours). The com-

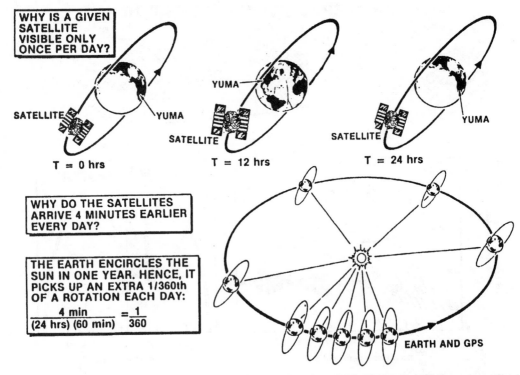

GPS ORBITAL GEOMETRY OVER A ROTATING EARTH

WHY IS A GIVEN SATELLITE VISIBLE ONLY ONCE PER DAY?

SATELLITE   YUMA   T = 0 hrs

YUMA   SATELLITE   T = 12 hrs

SATELLITE   YUMA   T = 24 hrs

WHY DO THE SATELLITES ARRIVE 4 MINUTES EARLIER EVERY DAY?

THE EARTH ENCIRCLES THE SUN IN ONE YEAR. HENCE, IT PICKS UP AN EXTRA 1/360th OF A ROTATION EACH DAY:

$$\frac{4 \text{ min}}{(24 \text{ hrs}) (60 \text{ min})} = \frac{1}{360}$$

EARTH AND GPS

Figure 10.7 The simple, geometrical sketches at the top of this figure would seem to show that each GPS satellite has a repeating ground-trace with a 24-hour period. However, as the more refined diagrams at the bottom of the figure clearly indicate, the earth travels around the sun approximately 1 degree per day. Thus, if a GPS satellite is overhead at noon today, it will be overhead at 4 minutes before noon tomorrow.

bination of these two fixed rotation rates produces a ground-trace repetition interval of approximately 24 hours.

The actual repetition interval is actually a little shorter than has been indicated so far. This is true because a slight adjustment is necessary to account for the fact that, while the satellite is whirling around the earth, the earth is whirling around the sun. During each 24-hour period the earth travels around the sun approximately 1 degree. The net result of these two simultaneous circular motions is that, if a particular GPS satellite was directly overhead at noon today, it will be overhead at the same spot tomorrow at 4 minutes before noon. The next day it will be directly overhead at 8 minutes before noon, and so on.

11

Precise Time Synchronization

Eighteenth century British sailors exhibited an almost haughty disdain for accurate navigation. When one of them was asked how to navigate a sailing ship from London to the New World, he replied: "Sail south until the butter melts, then turn right." For decades thereafter, Britannia ruled the waves, but her seamen paid for their lack of navigational expertise with precious ships and expensive cargos. Sometimes they paid with their own lives.

A special exhibit in the British Maritime Museum at Greenwich underscores some of the painful consequences of inaccurate navigation. In 1691, for instance, several ships of war were lost off Plymouth when the navigator mistook the Deadman for Berry Head. And in 1707 another devastating incident occurred when Sir Cloudsley Shovel was assigned to guide a flotilla from Gibraltar to the docks of London. After 12 days shrouded in heavy fog, he ran aground at the Scilly Islands. Four ships and 2,500 British seamen were lost.

These and a number of other similar disasters at sea motivated Parliament to establish the British Board of Longitude, a committee composed of the finest scientists of the day. They were charged with the responsibility of discovering some practical scheme for determining the locations of British ships on transoceanic voyages. In 1714 the Board offered a £20,000 prize to anyone who could provide them with a method for fixing a ship's position to within 30 nautical miles after six weeks at sea.

One promising possibility originally proposed by the Italian scientist Galileo would have required that navigators take precise sightings on the moons of Jupiter as they were eclipsed by the planet. If this technique had been adopted, special astronomical guides listing the predicted time for each eclipse would have been furnished to the captain of every flagship, or perhaps every ship in the British fleet. Galileo's elegant theory was entirely sound, but, unfortunately, its eighteenth century proponents were never

---

Using Wounded Dogs to Navigate Ships on the High Seas

Finding the latitude of a sailing ship can be surprisingly easy: sight the elevation of the Pole Star above the local horizon. Finding longitude turns out to be quite a bit harder because, as the earth rotates, the stars sweep across the sky 15 degrees every hour. A one-second timing error thus translates into a 0.25–nautical mile error in position. How is it possible to measure time onboard a ship at sea with sufficient accuracy to make two-dimensional navigation a practical enterprise?

One eighteenth century innovator, whose name has long since been forgotten, advocated the use of a special patent medicine said to involve some rather extraordinary properties. Unlike other popular nostrums of the day, "the Power of Sympathy," as its inventor, Sir Klenm Digby, called it, was applied not to the wound, but to the weapon that inflicted it. *The World of Mathematics*, a book published by Simon and Schuster, describes how this magical remedy was to be employed as an aid to maritime navigation:

"Before sailing, ever ship should be furnished with a wounded dog. A reliable observer on shore, equipped with a standard clock and a bandage from the dog's wound, would do the rest. Every hour on the dot, he would immerse the dog's bandage in a solution of the Power of Sympathy and the dog on shipboard would yelp the hour."

As far as we know, this intriguing method of navigation was never actually tested under realistic field conditions, so we have no convincing evidence that it would have worked as advertised.

---

able to devise a way to make the necessary observations under the rugged conditions existing at sea.

Another approach called for a series of "lightships" to be anchored along the principal shipping lanes on the North Atlantic. The crew of each lightship would fire luminous "star shells" at regular intervals timed to explode at an altitude of 6,400 feet. A ship in the area could calculate the distance to the nearest lightship by timing the duration between the visible flash and the sound of the exploding shell.

John Harrison's Marine Chronometer

Even before the dawning of the eighteenth century, the latitude of a maritime vessel was relatively easy to ascertain. At any location in the Northern Hemisphere, its latitude equals the elevation angle of the Pole Star. But determining its longitude has always been far more difficult because the earth's rotation causes the stars to sweep across the sky 15 degrees with every passing hour. A one-minute timing error thus translated into a 15–nautical mile error in longitudinal position. Unfortunately, measuring the time with sufficient accuracy aboard a rocking, rolling ship presented a forminable set of engineering problems.

In 1714, when the British Board of Longitude made its tantalizing announcement, a barely educated British cabinetmaker named John Harrison was perfectly poised to win the prize. Harrison had always been clever with his hands, and he had been blessed with a natural talent for repairing and building precision machinery. Moreover, when the British Board of Longitude announced its fabulously inviting proposition, John Harrison just happened to be a poor but energetic 21-year-old!

Flushed with the boundless enthusiasm of youth, he began to design and build a series of highly precise timekeeping devices. It took him almost 50 years of difficult labor, but in 1761 he was finally ready to claim the prize. Harrison's solution involved a new kind of shipboard timepiece, the marine chronometer, which was amazingly accurate for its day. Onboard a rolling ship, in nearly any kind of weather, it gained or lost, on average, only about one second per day.

Even by today's standards, Harrison's marine chronometer was a marvel of engineering design. He constructed certain parts of it from bimetallic strips to compensate for temperature changes; he used swiveling gimbal mounts to minimize the effects of wave-induced motions; and he rigged it with special mechanisms so that it would continue to keep accurate time while it was being wound.[1] Over a period of 47 years, Harrison built four different versions of the marine chronometer, all of which are, today, on display in Greenwich at the British Maritime Museum.

Unfortunately, by the time John Harrison managed to finish his fourth and final marine chronometer, he did not have enough strength left to stake his claim. So he persuaded his son, William, to travel from London to Jamaica to demonstrate its fabulously accurate navigational capabilities. During that entire six-week journey, the marine chronometer lost less than one minute. And, upon arrival at Jamaica, it helped fix the position of the ship to an accuracy of 20 nautical miles.

Disputes raged for years thereafter as to whether John Harrison should be declared the winner. At one point, members of The Board of Longitude insisted on confiscating his clever invention. They even tested it upside down, although Harrison had not designed it to keep accurate time in that unlikely method of operation. Eventually, through the intervention of royalty, John Harrison was awarded the entire £20,000 prize.

Celestial Navigation Techniques

The marine chronometer has, for decades, been used in conjunction with the sextant to fix longitudes and latitudes of vessels at sea. A sextant is an optical device that can be used to measure the elevation angle of any visible celestial body above the local horizon. While sighting planet or star through the optical train of the sextant, the navigator makes careful adjustments until the star's image is superimposed on the local horizon. A calibrated scale mounted on the side of the instrument then displays the elevation angle of the star.

---

[1]Once the marine chronometer was widely adopted for maritime navigation, any sailor who failed to wind it, when it was his assigned duty to do so, could be charged with a capital crime.

A precisely timed sextant sighting of this type fixes the position of the ship along a circular line of position lying on the spherical earth. By making a similar sighting on a second celestial body, with a different elevation angle, the navigator can construct a second circular line of position that will, generally speaking, intersect the first circle at two locations. He or she can then resolve the ambiguity either by having a fairly accurate estimate of the ship's position or by taking one more sextant sighting on a third celestial body.

Celestial navigation is still widely used by mariners all around the world, although its popularity is eroding as other more accurate and convenient navigation techniques pass into common use. Lewis and Clark used celestial navigation when they constructed accurate maps of the North American wilderness, and many Arctic explorers employed similar methods to guide the initial phases of their expeditions toward the North and the South Poles. The Apollo astronauts also relied on sextant sightings for a backup navigation system as they coasted silently through cislunar space. For these and many other applications of celestial navigation, precise time measurements are inevitably the key to achieving the desired accuracy and the desired confidence in the measured results.

A Short History of Time

Over the past one thousand years advancing technology has given us several generations of increasingly accurate clocks. Indeed, as the graph in Figure 11.1 demonstrates, today's best timekeeping devices are at least a trillion times (12 orders of magnitude) more stable and accurate than the finest clocks available 800 years ago. At the beginning of the tenth century, the most accurate timekeeping devices were water clocks and candle clocks, which, on the average, gained or lost approximately one hour per day. Balance clocks, which were widely adopted in the fourteenth century, kept time to within 15 minutes per day.

The next major advance in clockmaking technology was triggered by a simple observation by Galileo who, in 1651 (so the story goes) happened to wander into the church at the Leaning Tower of Pisa. Once inside, he noticed something that quickly captured his fancy: a candle suspended on the end of chain swinging in the breeze. Numerous other church-going Italians had witnessed the same thing hundreds of times before. But Galileo noticed something all of them had failed to recognize: The amount of time required for the candle to swing back and forth was independent of the swinging arc. When it traveled along a short arc it moved more slowly. When it traveled along a longer arc it moved faster to compensate. Galileo never used his clever pendulum principle to build a better clock, but he did suggest that others do so, and they were quick to follow that sound advice. Grandfather clocks, with their highly visible pendulums, are today's most obvious result.

THE HISTORICAL DEVELOPMENT OF ACCURATE CLOCKS

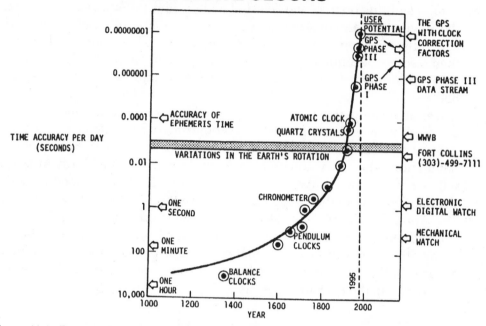

Figure 11.1   During the past 800 years timekeeping accuracies have improved by at least 12 orders of magnitude as innovative clockmaking technologies have been continuously introduced. In the twelfth century the best available timekeeping devices, candle clocks and water clocks, lost or gained 50 or 60 minutes during a typical day. Some of today's hydrogen masers would require several million years to gain or lose a single second. In the intervening centuries, pendulum clocks, marine chronometers, quartz crystal oscillators, and cesium atomic clocks have all, in turn, greatly improved mankind's ability to keep accurate time.

A well-built grandfather clock loses or gains perhaps 20 seconds in an average day.

Another important advancement came when, in 1761, after decades of labor, John Harrison managed to perfect his fourth marine chronometer, a precision shipboard timepiece that reduced timing errors to approximately one second per day. Thus, his device was just about as stable and accurate as a modern digital wristwatch that can be purchased for $30 in any large department store.

In the 1940s clocks driven by tiny quartz crystal oscillators raised time-keeping accuracies to impressive new levels of precision. A quartz crystal oscillator is a tiny slab of quartz machined to precise dimensions that oscillates at an amazingly regular frequency. Once quartz crystal oscillators had been perfected, they turned out to be more stable and accurate than the timing standard of the day, which was based on the earth's steady rate of rotation. Astronomers measured the relentless passage of time by making optical sightings at the zenith crossings of celestial bodies as they swept across the sky.

A few years later a new kind of official time standard was adopted, based on atomic clocks driven by the unvarying oscillation frequencies of cesium, rubidium, and hydrogen atoms. Voting networks that include the timing pulses from widely separated atomic clocks still serve as the global time standard for the Western world. Today's hydrogen masers are highly temperamental, but they are so stable and accurate that they would require millions of years to lose or gain a single second.

The Atomic Clocks Carried Aboard the Navstar Satellites

Only in the modern era of atomic clocks has timekeeping technology provided sufficient accuracy to allow the successful construction of the Navstar Global Positioning System. The evenly spaced timing pulses coming down from each Navstar satellite are generated by an atomic clock that contains no gears or cogs. Its extraordinary timekeeping abilities arise from the quantum mechanical behavior or certain specific atoms (cesium, rubidium, hydrogen), which tend to have a single outer-shell electron.

Cesium Atomic Clocks

Cesium atoms can exist in either of two principal states. In the high-energy state, the spin axis of the lone outer-shell electron is parallel to the spin axis of the atom's nucleus. In the low-energy state, the electron spins in an antiparallel direction. For cesium, the energy difference between the two spin states corresponds to an electromagnetic frequency of 9,192,631,770 cycles per second. Thus, when a cloud of cesium gas is struck by a radio wave oscillating near that particular frequency, some of the low-energy atoms will absorb one quanta of energy and, consequently, their outer-shell electron will flip over and begin spinning in the opposite direction. The closer the trigger frequency can be adjusted to 9,192,631,770 cycles per second, the more low-energy electrons will reverse their direction of spin.

The heart of a cesium atomic clock is a voltage-controlled crystal oscillator—a small vibrating slab of quartz similar to the one that hums inside a digital watch. Small variations in the voltage feeding a voltage-controlled crystal oscillator create corresponding variations in its oscillation frequency. Any necessary adjustments are handled by a feedback control loop consisting of a cesium atomic clock wrapped around the quartz crystal oscillator.

A schematic diagram of the cesium atomic clocks carried onboard the GPS satellites is sketched in Figure 11.2. First solid cesium is vaporized at 100°C, and then it is routed through a collimator to form a steady stream of cesium gas, which, in its natural state, consists of an equal mixture of high-energy and low-energy atoms.

THE CESIUM ATOMIC CLOCKS CARRIED ONBOARD THE NAVSTAR SATELLITES

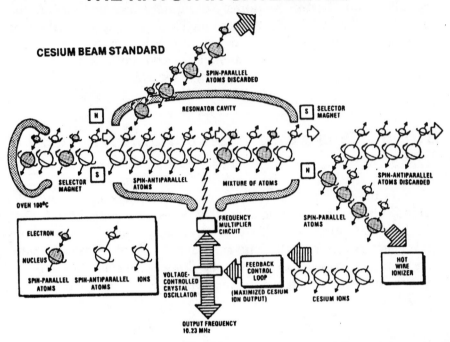

Figure 11.2 The low-energy atoms floating around inside the resonating chamber of this cesium atomic clock are hit with a radio wave as close as possible to 9,192,631,770 oscillations per second. Depending on the accuracy of that trigger frequency, larger or smaller numbers of the low-energy atoms will absorb one quanta of energy to become high-energy atoms— which are subsequently converted into cesium ions by the hot-wire ionizer (bottom right). The resulting ion current automatically adjusts the frequency of the quartz crystal oscillator, which, in turn, creates more timing pulses and precisely controlled electromagnetic waves.

A selector magnet is then used to separate the cesium atoms into two separate streams. The high-energy atoms are discarded: The low-energy atoms are deflected into a resonating cavity with precisely machined dimensions where they are hit with radio waves generated by a voltage-controlled crystal oscillator coupled to a solid-state frequency multiplier circuit. The closer the trigger frequency is to 9,192,631,770 oscillations per second, the more outer shell electrons will be inverted to produce high-energy cesium atoms.

When the atoms emerge from the resonating cavity, they are again sorted by a selector magnet into two separate streams. This time the low-energy atoms are discarded. The high-energy atoms are deflected onto a hot-wire ionizer, which strips off their outer-shell electrons to produce a stream of cesium ions. The resulting current is then routed into a feedback control loop connected to the voltage-controlled crystal oscillator whose oscillation frequency is constantly adjusted to produce new radio waves.

By adjusting the frequency to maximize the ion current and dithering the oscillator to make its frequency straddle the desired value of 9,192,631,770 oscillations per second, the frequency stability of the quartz crystal oscillator can be maintained to within 1 part in 5 billion. Thus, the feedback control loop just described stabilizes the frequency of the quartz crystal by a factor of 10,000 or so, compared with a free-running quartz crystal with similar design characteristics.

Rubidium Atomic Clocks

The rubidium atomic clocks carried onboard the GPS satellites are, in many respects, similar to the cesium atomic clocks, but there are also important differences in their design. For one thing, the rubidium atoms are not used up while the device is keeping time. Instead, the atoms reside permanently inside the resonating chamber. The sensing mechanisms that monitor and adjust the clock's stability are also based on distinctly different scientific principles.

As the rubidium atoms linger inside the resonating chamber, they are impacted with electromagnetic waves whose oscillation frequencies are as close as possible to 6,834,682,613 oscillations per second. As the transmission frequency is adjusted closer and closer to that precise target value, larger numbers of rubidium atoms will absorb one quanta of energy. When they do, their spin-states automatically reverse to convert them from low-energy to high-energy atoms.

The rubidium atomic clock converges toward the desired frequency through a feedback control loop whose status is continuously evaluated by shining the beam of a rubidium lamp through the resonating chamber. The gas inside the chamber becomes *more* or *less* opaque to rubidium light, depending on how many of the rubidium atoms inside have been success-fully inverted. The intensity of the rubidium light passing through the chamber is measured by a photo detector, similar to the electric eye in an instamatic camera.

The output from the photo detector is fed into a set of solid-state inte-grated circuits rigged to make subtle and continuous adjustments to the frequency of the voltage-controlled crystal oscillator. Pulses from the crystal oscillator, which vibrates at 5 million oscillations per second, are used in generating the evenly spaced C/A- and P-code pulses broadcast by the satellite. A portion of the output from the voltage-controlled crystal oscilla-tor is also fed into a set of frequency multiplier circuits, which generate the desired 6,834,682,613 oscillation-per-second frequency, which is, in turn, routed into the atomic clock's resonating chamber.

Developing Atomic Clocks Light Enough to Travel Into Space

When the architecture for the Navstar navigation system was first being selected, many experts argued convincingly that the atomic clocks should remain firmly planted on the ground. The C/A- and P-code pulse trains, they believed, should be sent up to the satellites through radio links for rebroadcast back to the users down below. This contentious position was quite defensible because all available atomic clocks were big and heavy, power hungry, and extremely temperamental.

The best available cesium atomic clocks operated by the National Bureau of Standards, for instance, were larger than a household deep freeze, and they had to be tended by a fretful army of highly trained technicians. However, emerging technology soon produced much smaller and far more dependable atomic clocks. After years of intellectual struggle, the cesium and rubidium atomic clocks onboard the Navstar satellites have turned out to be surprisingly small and compact. They also consume moderate quantities of electricity and can operate for several years without failure. The rubidium clocks carried aboard the Navstar satellites are roughly the same size as a car battery. Each one weighs about 15 pounds. The cesium atomic clocks are a little bigger. They weigh 30 pounds each.

The earliest Navstar GPS performance specifications called for atomic clocks with fractional frequency stabilities of 1 part in a trillion (10^{-12}). The fractional frequency stability of an atomic clock can be defined as the 1σ error pulse-to-pulse divided by the duration between pulses. An atomic clock with a fractional frequency stability of 1 part in a trillion is capable of keeping time to within one second over an interval of 30,000 years.

Although this performance specification may seem rather stringent, the first few spaceborne atomic clocks were two to five times more stable than required. Consequently, the specification goal was eventually raised to 2×10^{-13}.

The Navstar clocks have turned out to be surprisingly accurate and stable, but clock *reliability* problems plagued the first few GPS satellites. On the average, only five on-orbit months went by before a satellite component failure occurred. Almost always it was an atomic clock component that failed. With intensive design efforts, these problems were eventually brought under control so that, today, the probability that at least one of the four atomic clocks on a Block II satellite will still be operating at the end of its 7.5-year mission is estimated to be 99.44 percent.

The Growing Need for Precise Time Synchronization

The Navstar satellites were designed primarily for accurate navigation on a global scale, but they are also providing unprecedented opportunities for the global distribution of precise time. With reasonable care, distant atomic clocks can be synchronized to within 1/30,000,000 of a second using the signals from the Navstar satellites.

Even in the most demanding social situations, ordinary individuals almost never need to coordinate their activities to within 1/30,000,000 of a second. Synchronization accuracies in that rather stringent accuracy regime are required only when machines are talking to machines. Of course, in the modern world machines talk to one another in surprisingly large numbers, thus creating growing demands for precise time synchronization. Computer networks, for instance, operate much more efficiently if they are precisely synchronized. With sloppier time synchronization, larger "guard bands" are required to make sure the signals streaming between the various machines do not overlap.

Precise time synchronization also increases the efficiency of communication networks for much the same reason, especially if they span continent-wide distances and, most especially, if they include radio frequency transmissions through outer space. Bell Telephone's communication networks, for example, include at least 20 cesium atomic clocks to help foster efficient interconnections and message interleaving. Likewise, the Digital Equipment Corporation is keenly interested in gaining assured access to the precise GPS timing signals to enhance the efficiency of their local area computer networks.

Astronomy and terrestrial surveying require precise time synchronization, and so do ground-based and space-based navigation transmitters. Each Loran C transmitter is equipped with a pair of cesium atomic clocks. And the Omega transmitters rely on cesium atomic clocks, too. Satellite tracking and military surveillance can also benefit from precise time synchronization, as can certain types of military encryption.

Using GPS Time Synchronization to Help Make False-color Images of the Nightime Sky

A single, large radio telescope operating in isolation provides surprisingly poor resolution. The 135-foot radio telescope at Owens Valley, California, for instance, can resolve a dime-sized radio frequency source from a distance of only about 10 feet. By comparison, the unaided human eye can distinguish a dime-sized spot 40 feet away. When several radio telescopes are connected together to form a synchronized interferometry network, however, their resolution can be greatly increased. Such a network employs sophisticated time synchronization methods and computer processing techniques to combine the electromagnetic radiation gathered from multiple radio telescopes. The result is a "synthetic aperture" array that uses the natural rotation of the earth over an interval of several hours to provide systematic scanning.

When three or more radio telescopes are incorporated into such an array, the biggest source of image distortion can be virtually eliminated: ripples caused by atmospheric inhomogeneities. Once this has been accomplished, the resolution turns out to be 500 times better than what astronomers can achieve with Mount Palomar's giant 200-inch telescope.

Today's requirement for 300-nanosecond accuracies for synthetic aperture arrays can be met by linking the various radio telescopes together with hard-wired circuits or radio links. Tomorrow's more ambitious networks, which may be built with much longer baselines aboard spaceborne platforms, may have ten times better resolution, but they will require synchronization to within 30 nanoseconds. Fortunately, the common-view mode using the GPS satellites to measure time can easily provide the required precision.

Time Sync Methodologies

The barcharts in Figure 11.3 compare the accuracies for some of today's most popular methods of time synchronization. Notice that the synchronization methods listed are divided into two separate categories: those that provide local area coverage (such as Fort Collins telephone) and those that provide global coverage (such as the Transit navigation satellites). Notice also that the vertical scale has logarithmic spacings ranging over nine orders of magnitude. Each major division corresponds to a factor of 1,000 improvement in time synchronization accuracy, so a slight difference between the heights of two adjacent bars corresponds to a big improvement in the time synchronization accuracy that results. The four methods listed on the right-hand side of the figure are the most interesting because they provide global time synchronization services in direct competition with the GPS.

Under normal conditions, radio telephone can provide synchronization capabilities accurate to within one-thousandth of a second. A timing accuracy improvement of two orders of magnitude can be achieved by using, instead, the timing pulses from the Transit navigation satellites as they sweep across the sky. Portable atomic clock trips can provide another order-of-magnitude improvement, thus yielding time synchronization to within one-millionth of a second.

SYNCHRONIZATION CAPABILITIES OF VARIOUS SYSTEMS

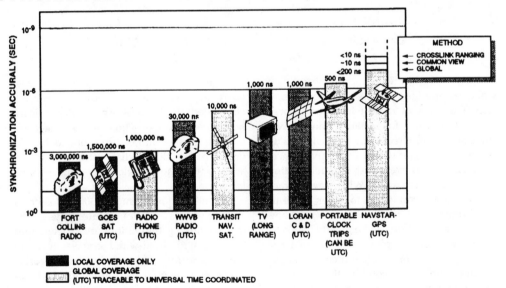

Figure 11.3 Eight methods for synchronizing widely separated atomic clocks are depicted in this figure. Of the ones that provide global coverage, the Navstar GPS has, by far, the most accurate time synchronization capabilities. Depending on the specific techniques being used, the Navstar signals can yield timing corrections to within a few nanoseconds keyed to the global time standard maintained in Paris, France.

A portable atomic clock is a small but precise timing device carried from one large, stationary atomic clock to another. The technicians first synchronize their portable atomic clock with respect to the atomic clock at some reference center, such as the National Bureau of Standards in Washington, D.C. Then they board a commercial jetliner and fly with the portable atomic clock to the second stationary clock in Paris, for instance, where they perform a second synchronization. Finally, they return the portable atomic clock to the original site, where they can perform one final synchronization to check the portable clock's drift rate during transit.

For each portable atomic clock trip, the technicians must purchase two and a half first-class airline tickets. The portable atomic clock, which must be cable-connected to the jetliner's power system, occupies a seat, but it does not eat airline food, so it flies at half fare. A few years ago a special exhibit on atomic clocks at the Smithsonian Institution displayed one of the half-fare airline tickets. The passenger's name on the ticket was "Mr. Cesium Clock."

Portable atomic clock trips have for years provided a rather glamorous profession for a few fortunate individuals. But, unfortunately, the necessity for this profession is quietly in decline. This is true because today's Navstar satellites reduce the time synchronization errors by at least one or two orders of magnitude compared with the finest available portable atomic clocks.

Fixing Time with the Navstar Signals

As Figure 11.4 indicates, the time synchronization techniques currently employed by the Navstar user community can be partitioned into four separate categories:

1. Absolute time synchronization
2. Clock fly-overs
3. The common view mode
4. The multi-satellite common view mode

With *absolute time synchronization* a special time sync receiver picks up the signals from a single GPS satellite as it travels overhead. With this simple technique, synchronization accuracies of about 100 nanoseconds can be achieved when selective availability is not being implemented. With selective availability, the worst-case C/A-code time synchronization error may be as large as 300 nanoseconds.

When *clock fly-overs* are used, the satellite swings up over the two sites, one after the other. As it passes over each site, careful clock synchronization operations are performed to determine timing errors for each ground-based atomic clock relative to the satellite. The clock fly-over technique is similar to a portable atomic clock trip, but the satellite travels across continent-wide distances much faster than any commercial jetliner, so the timing errors turn

TIME TRANSFER VIA NAVSTAR-GPS SATELLITE

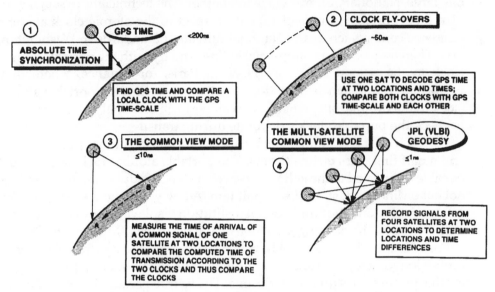

Figure 11.4 Four promising methods of precise time synchronization have been developed by the users of the Navstar GPS. All are practical and effective, but the common view mode depicted in the lower left-hand corner has become increasingly popular among timekeeping experts around the world. In the common view mode, both clock sites on the ground must have direct, line-of-sight access to the same Navstar satellite at the same time. By observing their time offsets with respect to the GPS satellite and communicating the results to one another, the timing errors between their clocks can be reduced to only about 10 nanoseconds.

out to be much smaller. Under typical conditions, the clock fly-over mode yields a synchronization error of around 50 nanoseconds.

The *common view mode* can be implemented when two distant sites have direct, line-of-sight access to the same GPS satellite at the same time. By synchronizing their atomic clocks with respect to the satellite clock, and then comparing their clock offsets with respect to one another, clock experts can consistently achieve time synchronization errors of 10 nanoseconds or less.

The *multi-satellite common view mode* involves four or more satellites that are being observed simultaneously from the two different clock sites. With rather extensive computer processing, this approach can provide time synchronization accuracies clearly superior to the accuracies that can be achieved by using the conventional common view mode. The time synchronization experts at JPL's deep-space network, for example, believe they may be able to achieve time synchronization errors as small as 1 nanosecond.

For time synchronization applications involving an atomic clock on a moving platform or one at an unknown location, a complete four-satellite navigation solution must be carried out in conjunction with the time synchronization. In this rather demanding case, the user set must solve for its three position coordinates U_x, U_y, U_z, together with its clock bias error, C_B, as shown in Figure 11.5. This, of course, requires simultaneous access to at least four Navstar satellites.

For stationary users at known locations, the C/A-code signal from a single satellite can provide accurate time synchronization. Access to four satellites is not required, because the user's position coordinates U_x, U_y, U_z are already known—or can be obtained in advance using GPS surveying equipment. The substantial sales of time synchronization receivers early in the GPS program came about primarily because this rather non demanding requirement for single-satellite access when the number of satellites in the constellation was quite small.

Lightweight Hydrogen Masers for Tomorrow's Navstar Satellites

Improved clock stability and timing accuracy could be achieved by installing compact hydrogen masers aboard the Navstar satellites. Only a few years ago hydrogen masers were far too big, heavy, and power hungry for

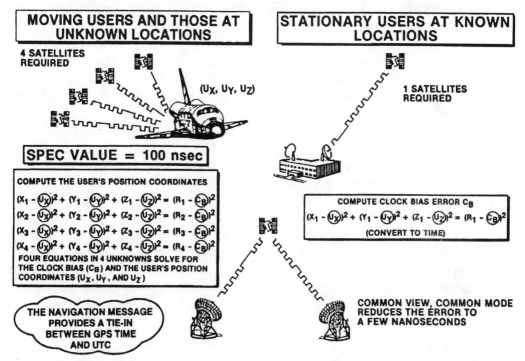

Figure 11.5 For moving atomic clocks and for those at unknown locations, access to four or more Navstar satellites is required in order to achieve precise time synchronization. This four-satellite requirement stems from the fact that the receiver must solve for both its position and the exact time. For atomic clocks at known, stationary locations, access to a single GPS satellite during the time synchronization interval will usually suffice.

the Navstar mission, but careful engineering techniques have quietly reduced their size, weight, and power consumption. The atomic clock experts at the Hughes Research Center in Malibu, California, have managed to develop a compact hydrogen maser that is only 18 inches long, weighs less than 50 pounds, and draws only about 45 watts of electrical power. Well-designed hydrogen masers are 50 to 100 times more stable than the cesium atomic clocks carried onboard the Navstar satellites.

Tests at the Hughes Research Laboratory have demonstrated fractional frequencies stabilities as low as 2×10^{-15}. The timing uncertainty associated with such a clock is only about 1 second every 7 million years. Such stability is quite hard to visualize, but imagine this: If the first cave man had had such a clock strapped to his wrist on the day he was born, it would not yet have lost a single second.

The operating principles of the compact hydrogen maser are quite esoteric, but fairly easy to understand. The process starts when ordinary molecular hydrogen is routed through an electrical discharge tube. This converts it from molecular hydrogen involving connected pairs of hydrogen atoms into atomic hydrogen, which is composed of monotomic atoms, each of which has a single orbiting electron. The atoms in monotomic hydrogen come in two principal species: "spin-up atoms" and "spin-down atoms." For *spin-up* atoms the spin axis of the electron points upward, away from the nucleus. For *spin-down* atoms the electron's spin axis points downward toward the nucleus (see Figure 11.6).

This nearly equal mixture of spin-up and spin-down hydrogen atoms is then routed through a hexapole selector magnet, which discards the spin-down atoms. When they are discarded, the spin-down atoms are collected on a "gettering" material—a flypaper-like substance with an affinity for hydrogen atoms.

The spin-up atoms are routed into a glass bulb that is coated on the inside with a thin, molecular layer of teflon. Once inside, each spin-up atom bounces off the smooth walls of the teflon-coated bulb, on the average, about 100,000 times before it experiences a spontaneous state-change to become a spin-down atom. Whenever such a state-change occurs, a burst of microwave energy is emitted, a small portion of which is picked up by the microwave radiation antennas protruding into the resonating cavity.

The compact hydrogen masers built at the Hughes Research Center are approximately 20 inches long. They weigh about 43 pounds. The full-sized devices built by the experts at other institutions, such as the Cambridge Astrophysical Observatory, are even more stable, but they are also quite a bit bigger and heavier.

Crosslink Ranging Techniques

Hydrogen masers have been studied extensively for possible use aboard future Navstar satellites, in part because they could theoretically make the

HUGHES AIRCRAFT'S HYDROGEN MASER

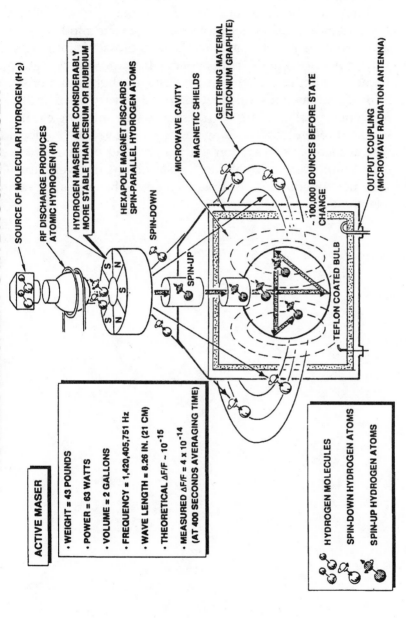

ACTIVE MASER

- WEIGHT = 43 POUNDS
- POWER = 63 WATTS
- VOLUME = 2 GALLONS
- FREQUENCY = 1,420,405,751 Hz
- WAVE LENGTH = 8.26 IN. (21 CM)
- THEORETICAL $\Delta F/F \sim 10^{-15}$
- MEASURED $\Delta F/F = 4 \times 10^{-14}$
  (AT 400 SECONDS AVERAGING TIME)

SOURCE OF MOLECULAR HYDROGEN (H$_2$)

RF DISCHARGE PRODUCES
ATOMIC HYDROGEN (H)

HYDROGEN MASERS ARE CONSIDERABLY
MORE STABLE THAN CESIUM OR RUBIDIUM

HEXAPOLE MAGNET DISCARDS
SPIN-PARALLEL HYDROGEN ATOMS

SPIN-DOWN

MICROWAVE CAVITY

MAGNETIC SHIELDS

GETTERING MATERIAL
(ZIRCONIUM GRAPHITE)

SPIN-UP

~ 100,000 BOUNCES BEFORE STATE
CHANGE

TEFLON COATED BULB

OUTPUT COUPLING
(MICROWAVE RADIATION ANTENNA)

HYDROGEN MOLECULES

SPIN-DOWN HYDROGEN ATOMS

SPIN-UP HYDROGEN ATOMS

Figure 11.6 In this compact hydrogen maser, a stream of spin-up hydrogen atoms is created by routing ordinary molecular hydrogen into an RF (radio frequency) discharge tube, and then sending the resulting atomic hydrogen through a hexapole selector magnet. When it enters the teflon-coated bulb, each spin-up hydrogen atom ricochets, on the average, about 100,000 times before it spontaneously inverts to produce a spin-down atom. This state change is accompanied by a burst of microwave energy that is picked up by the two microwave radiation antennas protruding through the walls of the resonating cavity.

163

satellites more autonomous—less dependent on frequent updates from the ground. However, an alternate approach called *crosslink ranging* promises to achieve similar results at a substantially lower cost. The crosslink ranging system, which is being implemented for use aboard the GPS Block IIR satellites, equips each satellite to send precisely modulated timing pulses to one another using their normal UHV crosslinking system. With properly modulated crosslinks, the GPS satellites can mutually synchronize their atomic clocks and measure their locations relative to one another.

If the earth always rotated at a constant rate, crosslink ranging could provide long-term autonomous independence from ground updates. Unfortunately, the rotation rate of the earth varies in slightly unpredictable ways, so the Block IIR satellites still must be updated every few months. This fixes their current orbital locations with respect to the spinning earth.

Practical benefits of crosslink ranging include improved navigation accuracies, longer autonomous operation without ground updates, increased service life, and improved survivability. Substantial manpower reductions for control segment operations are also anticipated when crosslink ranging is fully implemented.

12

Digital Avionics and Air Traffic Control

Safely directing enormous swarms of airplanes between America's 6,000 airfields is one of the most costly and labor-intensive navigation problems any nation has ever had to face. A total fleet of 3,500 commercial jetliners serve America's major cities, but 250,000 smaller civil aviation aircraft, mostly privately owned, are also hurtling across North American skies.

Several hundred ground-based navigation transmitters and radar installations tended by 10,000 air traffic controllers, help foster safe and efficient airborne operations, but, even with such an elaborate and costly infrastructure, America's present system of air traffic control is often straining toward the rupture point by ever-expanding traffic loads.

The Sabreliner's Flight to the Paris Air Show

Early proposals for navigating airplanes using signals from the Navstar satellites were often greeted with chilly receptions or outright skepticism. But that was before July 1983, when executives from the Government Avionics Division of Rockwell International dramatically demonstrated some of the more promising benefits of Navstar navigation. At that time two Navstar receivers were used to guide a Sabreliner business jet from Rockwell's production facility at Cedar Rapids, Iowa, across the Atlantic to the Paris Air Show. All along the way Navstar satellites were used to fix the position of the business jet, whose short passenger list included an impartial observer from the popular aerospace magazine *Aviation Week and Space Technology*.

The pilot landed four times along the way at the intermediate locations marked in Figure 12.1. Frequent landings were necessary because, at that time, only five Navstar satellites were broadcasting accurate navigation signals. Consequently, the flight crew had to descend periodically to wait for the Navstar's skimpy little constellation to come back up over the horizon. After the pilot touched down at Le Bourget Field near Paris, he taxied blind to a presurveyed spot along the apron of the runway, ending up within 25 feet of his intended destination.

Two records were set during that historic Sabreliner mission:

1. It was the first ever to use satellite navigation all the way across the Atlantic.
2. It took three times longer than Charles Lindbergh required in 1927.

Two different types of receivers were carried onboard: a single-channel dual-frequency military receiver and a single-channel single-frequency civilian receiver. Throughout the mission, the solutions were constantly

SABRELINER FLIGHT TO THE PARIS AIR SHOW

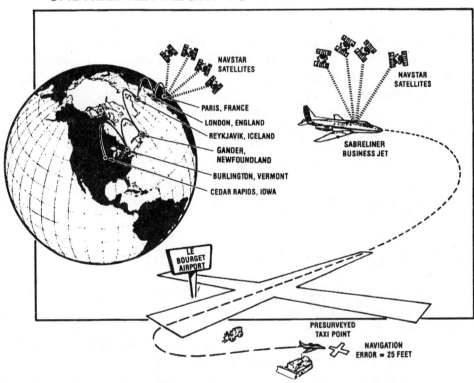

Figure 12.1 In 1983 a Sabreliner business jet was successfully vectored from Cedar Rapids, Iowa, to the Paris Air Show, using signals from the Navstar navigation satellites. The pilot landed five times along the way to wait for the small constellation of GPS satellites to come back up above the horizon. Once the business jet touched down at Le Bourget Field outside Paris, it was used repeatedly to fly local dignitaries in demonstration flights over the colorful French countryside.

recorded and compared in order to detect any navigation errors or systematic drift.

Four Major Concerns of the Federal Aviation Administration

Despite the apparent success of the Paris Air Show mission, cautious officials at the Federal Aviation Administration still had a number of reservations concerning the ultimate practicality of using Navstar navigation to vector and control airplanes cruising over America and throughout the rest of the world. Their worries centered around four specific technical and political issues:

1. Selective availability
2. User-set fees
3. Integrity-related failures
4. Continuous five-satellite coverage

These four major concerns will be discussed one by one in the next few subsections, together with any partial resolutions that have been adopted so far.

Selective Availability

Selective availability (degradation of accuracy) refers to the purposeful reduction in navigation accuracy imposed by the Department of Defense (DoD) to increase the positioning errors of any potential military applications conducted by unauthorized Navstar users. DoD policymakers initially wanted a civil 2σ positioning error of 1,650 feet (500 meters). But their counterparts at the Federal Aviation Administration argued forcefully for sufficient accuracy to handle a typical non-precision approach. In a series of meetings convened by the authors of the Federal Radionavigation Plan, these conflicting desires were carefully mediated. Once the dust had settled, the committee's recommendation called for a maximum 330 foot (100 meters) 2σ navigation error (95-percent probable) for the unauthorized users. This was just sufficient to support non-precision approach maneuvers for America's fleet of commercial and private planes.

User-Set Fees

The second major issue of concern to Federal Aviation Administration officials centered around the possible imposition of user-set fees. During the

Reagan Administration, the United States government began collecting fees for various government services (such as Coast Guard rescue) previously provided free to civilian users. Similar fees, it was believed, might be appropriately collected from users of the multi-billion dollar Navstar GPS who would, otherwise, benefit from its use for free. Accordingly, the U.S. Air Force was instructed to perform a series of studies on various possible fee collection methods.

In their final published study the researchers noted that they had been unable to devise any fee collection method that would, with any degree of certainty, provide positive revenues and economic benefits. In general, as Table 12.1 indicates, the various fee collection methods they evaluated might well have cost more dollars to implement than would have been collected. So user-set fees are no longer being considered in the Federal Radionavigation Plan.

Moreover, at a recent Future Air Navigation Systems (FANS) meeting, the U.S. government guaranteed that, for 10 years after the full constellation has been installed, GPS services will be donated to the world for free. Not to be outdone, the government of the Soviet Commonwealth has guaranteed cost-free access to the signals broadcast by the Glonass constellation for at least 15 years. Both governments are also guaranteeing a 330-foot or smaller horizontal navigation error at least 95 percent of the time to unauthorized (civilian) users.

Table 12.1 Standard positioning service user-charge concepts

Criteria	1 Card key	2 Software changes	3 One-time charge	4 Recurring charges	5 No charge
Encourages user acceptability and utilization	No	No	No	No	Yes
Provides revenues and economic benefits	Uncertain[1]	Uncertain[1]	Uncertain[1]	Uncertain[1]	No[2]
Easy to implement and maintain	No	No	Marginal	Marginal	Yes
Enhances operations and safety	No	No	Yes	Yes	Yes
Enforceable	Yes	Marginal	No[3]	No[3]	Yes
Protects national security	No	Yes	Yes	Yes	Yes
Consistant with current U.S. law and policies	No	No	No	No	Yes

[1]There is a probability of losing money.
[2]Money will not be lost.
[3]Enforceable on U.S. users and manufacturers only.

Integrity-related Failures

The third GPS issue of concern to administrators at the FAA deals with their long-standing requirement for reliable detection of integrity-related failures. An integrity-related failure occurs when a navigation system misleads its users by malfunctioning in some manner that produces inaccurate navigation, while providing no indication to the users that unacceptable errors have occurred.

Studies conducted by the Federal Aviation Administration center around three promising methods for monitoring the integrity of Navstar navigation:

1. Monitoring aboard the satellites
2. Monitoring aboard the airplanes
3. Monitoring on the ground

Summaries highlighting the key features of these three methods of integrity monitoring are presented in Figure 12.2.

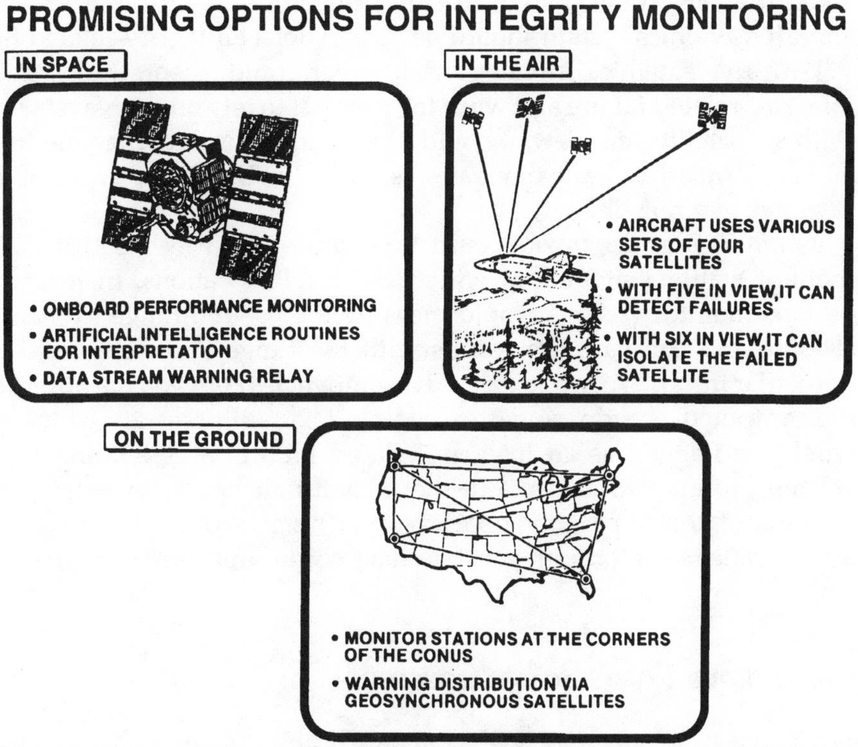

PROMISING OPTIONS FOR INTEGRITY MONITORING

Figure 12.2 In theory, the integrity of navigation signals streaming down from the GPS satellites could be monitored in space, in the air, or on the ground. Monitoring in space is costly and complicated, and it may not work as advertised. Monitoring in the air can be accomplished successfully only if coverage from six or more satellites can be guaranteed at all times. Monitoring on the ground requires widely separated ground-based facilities and continuously available satellite relay stations that can detect any satellite failures and issue integrity-related warnings within a few seconds after the failure has occurred.

Monitoring aboard the satellites could, in theory, be accomplished by analyzing and detecting all possible satellite failure modes, and then using artificial intelligence routines to interpret any anomalies that might occur. Warning messages could be transmitted separately, or they could be inserted in the 50-bit-per-second data stream coming down from each satellite. Unused data stream bits near the ends of subframes 2 and 4 could be made available for integrity warnings. However, engineering studies indicate that onboard integrity monitoring techniques would likely require complicated and expensive computer codes, most of which could never be tested under realistic field conditions. The possibility of missing some subtle failure mode worries responsible individuals and so does the inherent complexity of this proposed method for the onboard monitoring of the operating status of each satellite.

Monitoring aboard the airplanes will likely turn out to be the most convenient approach, assuming that a sufficiently large constellation of Navstar satellites can be launched into orbit to make it work properly. In this approach, the Navstar receiver carried aboard each aircraft would perform a series of navigation solutions, using various combinations of GPS satellites taken four at a time. By comparing the solutions with one another, the aircraft's avionics system should be able to detect integrity-related failures.

With five satellites in view, the receiver could, theoretically, detect an integrity-related failure and warn the user not to rely on the Navstar system. With six satellites in view, it could, theoretically, isolate the failed satellite and perform an accurate navigation solution, using the best four of the five satellites that remain.

Monitoring on the ground could be accomplished by a set of dedicated monitor stations situated at widely separated locations. In real time the various monitor stations would measure the pseudo-range to each visible satellite and compare the results with the range known to exist at that particular instant. Any unexpected discrepancies between the observed and the computed pseudo-ranges exceeding a certain prearranged magnitude would be flagged as an integrity-related failure, necessitating a prompt warning to all critical users in the relevant air space. In most proposed systems of this type, any necessary warnings would be relayed to the appropriate users through geostationary communication satellites.

Continuous Five-satellite Coverage

The overriding desire to achieve continuous five-satellite coverage presents additional problems for the Federal Aviation Administration. At one time budget pressures reduced the planned constellation to only 18 GPS satellites plus 3 active on-orbit spares. With that smaller constellation, coverage in some areas could drop to only four satellites for brief intervals at predictable times.

Continuous five-satellite coverage could be assured by increasing the number of satellites launched into the conventional 12-hour orbits. But another promising possibility would be to supplement the conventional 12-hour constellation with additional satellites launched into geostationary orbits. These additional GPS satellites could be designed to carry their atomic clocks onboard, or they could be designed as "orbiting transponders" that would relay C-/A- and/or P-code pulse sequences produced by atomic clock–equipped stations situated on the ground.

A constellation of 24 satellites, properly positioned in 12-hour orbits, can provide continuous 6-satellite coverage everywhere in the world. In theory, this would permit real-time integrity monitoring onboard each properly equipped aircraft. This approach would also yield more accurate navigation and a more robust Navstar constellation.

A variety of studies have been conducted comparing, contrasting, and refining the three major integrity monitoring techniques, but so far no definite selection has been made. However, a Navstar constellation consisting of 21 satellites plus three active on-orbit spares launched into 6 separate orbit planes has been approved by the DoD. This architecture was chosen, in part, because of forceful arguments from the Department of Transportation, whose experts realized that a larger constellation would be extremely beneficial to air-related users of Navstar navigation.

Using a Dedicated Constellation for Air Traffic Control

Many years ago Rockwell engineers at Anaheim, California, conducted a large and intensive study in which they examined the possible benefits of using orbiting satellites as the backbone of a clever new system of air traffic control. The advanced satellites they envisioned would provide accurate and continuous navigation services, together with two-way communication coverage for the continental United States, and beyond. Among other things, the satellites they envisioned were to provide dedicated voice channels linking the pilots with their air traffic controllers on the ground below.

At that time, the fundamental concepts for the Navstar Global Positioning System were still being formulated, so Navstar navigation was not incorporated into the Anaheim concept. Instead, the study engineers proposed a system with a new constellation of navigation/communication satellites dedicated to air traffic control. The constellation they proposed would have served the lower 48 states with a total of 15 satellites. Six of them were to be launched into geostationary orbits south of the United States with 20-degree longitudinal spacings. The other nine were to be launched into inclined elliptical 24-hour orbits, where they would trace out long, looping figure-8 ground tracks ranging between 80 degrees North and 80 degrees South latitude. The ground trace geometry and some of the other key characteristics of the Anaheim constellation are highlighted in Figure 12.3.

Three new regional air traffic control centers were to serve the continental United States: a mid-continent center located at Forth Worth, Texas, supple-

New Navigation Concept Arises out of Tragedy

In 1978 a new space-based navigation system, the Geostar, was born, with trip-hammer suddenness, when two airplanes slapped together high over the emerald blue swimming pools of sunny Southern California. "In clear skies at San Diego, the nation witnessed a horrible air tragedy," wrote Rob Stoddard in *Space Communications*. "Despite good weather conditions and state-of-the-art communications, a small private plane collided in midair with a crowded 727 commercial jetliner, sending scores of people plunging to their deaths." Dr. Gerald O'Neill of Princeton University, himself a frequent air traveler, was shattered by the early descriptions of the incident—even before he learned that a close friend was aboard the bigger plane. "There had to be a way to avoid this type of tragedy in the future," he later told a reporter.

By funneling his grief into constructive channels and using his considerable scientific expertise, O'Neill put together the concept for a satellite-based radionavigation system that provides precise position coordinates and allows its users to exchange simple "telegram messages" anywhere within the continental United States.

Unlike the Navstar receivers, which use computer processing techniques to perform their own navigation solutions, Geostar's receivers are more akin to Dick Tracy two-way wrist radios. When a Geostar receiver needs to know its position, interrogation pulses are automatically transmitted to all three of the geosynchronous satellites, which immediately relay the request to a centrally located computer sitting on the ground. The computer performs the navigation solution, and then sends the results back to the user through one of the three geosynchronous satellites. For a small fee, the same two-way communication channels can be used to relay short "telegram messages" between two or more Geostar subscribers.

Geostar's two-way communication links are more complicated than the ones that serve the Navstar users. But the receivers themselves are much simpler, and, for the mobile population O'Neill and his colleagues chose to serve, real-time communications were often more valuable than precise navigation.

By December 1985 Geostar had received firm orders for at least 5,000 transceivers, each secured by a 5-percent deposit against the full price of $2,900 each. In the early 1990s, when all three dedicated satellites were slated to become operational, Geostar's officials had hoped to have more sophisticated "pocket-size" receivers in production at a cost of about $500 each. Unfortunately, to create more aggressive competition, government regulations were changed. O'Neill's basic concept was technically sound, but, it may have been too advanced for its time. By 1991 Geostar's managers had run into serious difficulty with their company's financiers. Soon they were forced to fold up the business.

mented by two smaller ones in Florida and California. A total bandwidth of 101 megahertz was required for the system, which relied exclusively on conventional voice communications.

An Alternative Architecture Using the GPS

An alternative constellation architecture that takes advantage of the existing 12-hour GPS satellites supplemented by three more orbiting at geosynchronous altitudes is sketched in Figure 12.4. The three geosynchronous satellites provide C/A- and P-code navigation signals with triply redundant overlapping coverage for the continental United States. The supplemental geosynchronous satellites also act as communication nodes, linking air traffic control centers with the airplanes flying over the North American continent.

CONSTELLATION ARCHITECTURE OF THE ANAHEIM CONCEPT

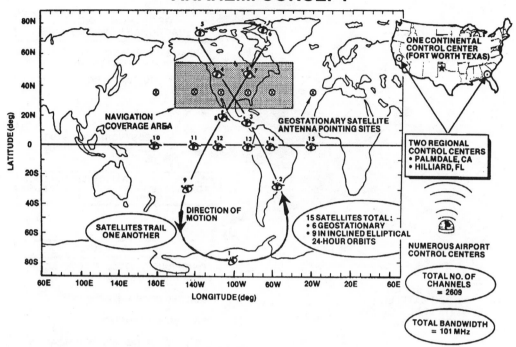

Figure 12.3 This early concept for a space-based system of air traffic control employs six geostationary satellites plus nine more in 24-hour, inclined elliptical orbits. Each of the 15 satellites in the constellation would be rigged to handle both communications and navigation to link the airplanes with the three regional air traffic control centers in Texas, Florida, and California. Tomorrow's systems of air traffic control will likely incorporate the continuous navigation coverage provided by the Navstar satellites and, perhaps, the Soviet Glonass satellites, too.

Advanced computer processing techniques, three-dimensional video displays, and voice synthesizer chips all combine to make this modern system considerably more robust and efficient than today's patchwork system or the dedicated space-based system sketched in Figure 12.3.

When the new geosynchronous satellites are coupled with the ones in the conventional 24-satellite GPS constellation, continuous 9-satellite coverage is provided for the lower 48 states. The new architecture also yields improved coverage geometry, reduced navigation errors, more precise time synchronization, and powerful new communication links to help reduce the dangers that will inevitably lurk in tomorrow's overcrowded skies.

Every airplane (excluding hang gliders and ultralights) will be equipped with a simple Navstar receiver. During flight, that receiver will determine the airplane's position, velocity, and the exact time, using the signals streaming down from the expanded constellation of Navstar satellites. Every few seconds the avionics system relays an appropriate version of its state vector to one of the three geostationary satellites hovering overhead.

ALTERNATE GPS CONSTELLATION ARCHITECTURE FOR THE FUTURE

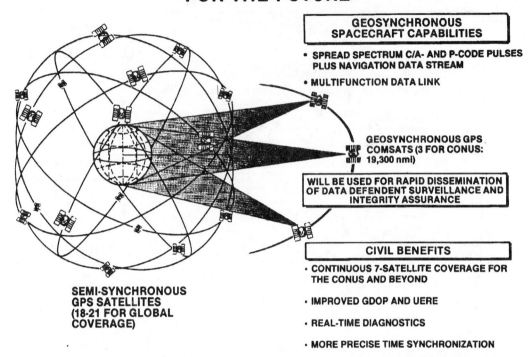

GEOSYNCHRONOUS SPACECRAFT CAPABILITIES

- SPREAD SPECTRUM C/A- AND P-CODE PULSES PLUS NAVIGATION DATA STREAM
- MULTIFUNCTION DATA LINK

GEOSYNCHRONOUS GPS COMSATS (3 FOR CONUS: 19,300 nmi)

WILL BE USED FOR RAPID DISSEMINATION OF DATA DEFENDENT SURVEILLANCE AND INTEGRITY ASSURANCE

CIVIL BENEFITS

- CONTINUOUS 7-SATELLITE COVERAGE FOR THE CONUS AND BEYOND
- IMPROVED GDOP AND UERE
- REAL-TIME DIAGNOSTICS
- MORE PRECISE TIME SYNCHRONIZATION

SEMI-SYNCHRONOUS GPS SATELLITES (18-21 FOR GLOBAL COVERAGE)

Figure 12.4 This promising concept for tomorrow's spaceborne system of air traffic control would incorporate the conventional 12-hour GPS satellites plus additional satellites at the geostationary altitude rigged to provide both navigation and communication. Three geostationary satellites would yield triply redundant coverage for the continental United States. Two to four others could provide essentially global coverage. With digital processing techniques, touch screen technology, advanced computer graphics, and voice synthesizer chips, this system could yield clear and unambiguous air traffic control services with several times the capacity provided by today's labor-intensive, ground-based position-fixing techniques.

The appropriate geostationary satellite picks up this state-vector information and broadcasts it back down again to blanket the continental United States with continuous information defining the current location and status of every aircraft flying over American airspace so that the air traffic controllers and the pilots of other, nearby airplanes, can display it on their own video screens.

Comparisons Between Geosynchronous and Semi-synchronous Constellations

Comparisons between the relative benefits obtained from placing the communication satellites in geosynchronous orbits or the conventional 12-hour

GPS orbits are presented in Table 12.2. Notice that the major *disadvantages* of the geosynchronous option include:

- More satellite power required
- Larger antenna apertures
- Difficulty in communication with airplanes at high northern latitudes due to low grazing angles
- Geosynchronous frequency and orbital slot assignment limitations

In addition, a larger launch vehicle would be required for the geosynchronous option because a heavier, more powerful satellite must be carried up to a higher altitude.

The geosynchronous option adds 400 pounds and 500 watts to the conventional Block II satellite. Thus, a typical geosynchronous GPS would experience a weight increase from about 2,000 to 2,400 pounds, and its power would increase from about 700 to 1,200 watts.

The McDonnell Douglas Delta II medium launch vehicle is not powerful enough to boost the heavier version of the Block II satellite to the higher geosynchronous orbit, but the Atlas-Centaur has abundant performance to handle the launch.

Piggyback Geosynchronous Payloads

INMARSAT's engineers are planning to launch piggyback payloads aboard their geosynchronous satellites capable of providing L-band–compatible GPS-type signals. The extra cost of the flight hardware capable of providing this valuable service is estimated to be only about 5 percent the cost of building and launching a conventional GPS satellite from scratch. INMARSAT's C/A-code pulse sequences would be generated on the ground for relay through conventional geosynchronous communication satellites.

Table 12.2 Relative advantages of supplemental geosynchronous air traffic control satellites

Advantages of the geosynchronous option	Disadvantages of the geosynchronous option
Viewing angle fixed geodetically	More satellite power required
Minimal antenna steering required	Larger aperture/gain required
Fixed communication node requires no switching of satellites	European/high-latitude coverage is difficult due to low elevation angle
More benign radiation environment	Geosynchronous slot assignment limitations
Less Doppler/range variation in receivers	Larger launch vehicle required

More elaborate versions of their piggyback payloads might eventually be able to provide communication services for air traffic controllers, thus reducing the cost of the proposed air traffic control system to an easily affordable level. Contracts for the INMARSAT satellites have already been released to the prime contractor, General Electric, with the GPS option included in the contract price. Four satellites with the desired capabilities are already under contract, with five more under option.

The Autoland System Test Results

An elaborate automatic landing (Autoland) system has been tested repeatedly to show that, with the proper supplementary equipment, GPS navigation can rival the landing accuracy provided by the new Microwave Landing System (MLS) now planned for installation at hundreds of large airports throughout the Western world. The Autoland System relies on differential navigation techniques, supplemented by a highly accurate radar altimeter installed aboard each plane.

Test engineers from NASA and Honeywell have demonstrated the Autoland System at Virginia's Wallops Island Test Facility. In their test series, the automated system was installed in a modified Boeing 707 aircraft equipped with two separate cockpits, one of which has no outside windows. During 36 separate landing tests, the crew in the concealed cockpit used only their instruments to land the plane.

Many aerospace experts are convinced that the Navstar GPS and similar spaceborne systems will eventually simplify aircraft navigation and help promote safer and more efficient air traffic control. If Navstar navigation can capture even a small fraction of this vast and expanding market, high-end receiver prices could quietly descend to much more affordable levels, thus encouraging the widespread adoption of Navstar navigation into even larger markets, such as private boating, global tourism, and motor vehicle transportation.

13

Geodetic Surveying and Satellite Positioning

In 1988 a team of surveyors used the signals from the Navstar satellites to reestablish the locations of 250,000 landmarks sprinkled across the United States. According to one early press report, their space-age measurements caused the research team to "move the Washington Monument 94.5 feet to the northwest!" And during that same surveying campaign, they moved the Empire State Building 120.5 feet to the northeast, and they repositioned Chicago's Sears Tower 90.1 feet to the northwest.

In reality, of course, the Navstar satellites do not give anyone the power to move large, imposing structures, but the precise signals they broadcast do provide our geodetic experts with amazingly accurate and convenient position-fixing capabilities that have been quietly revolutionizing today's surveying profession. Someday, soon, the deed to your house may be specified in GPS coordinates.

Surveying with a GPS receiver entails a number of critical advantages over classical ground-based methods for pinpointing the locations of widely scattered landmarks on the earth's undulating surface. For one thing, *inter-visibility* between benchmarks is not required. Navstar receivers positioned at surveyors benchmarks often have access to the signals from the GPS satellites sailing overhead even though they may not be within sight of one another. This can be especially important in tree-shrouded areas, such as the dense rainforests of Indonesia and Brazil. In such cluttered conditions, conventional surveying teams sometimes spend hours erecting big, portable towers at each site to achieve the required intervisibility high above the forest canopy. When it is time to move on, they tear the towers down one by one and lug their girders to different locations, and then build them back up again.

GPS surveying is advantageous because it is essentially weather-independent, and because it permits convenient and accurate day-night operations. With carrier-aided navigation techniques, site-to-site positioning errors as small as a quarter of an inch can sometimes be achieved.

The signals from the space-based Transit Navigation System have been used for many years to aid specialized terrestrial surveying operations. Unfortunately, Transit surveying suffers from a number of practical limitations as compared with similar operations using the GPS. A Transit satellite, for instance, climbs up above the horizon, on average, only every hour or so compared with continuous GPS satellite observations. Moreover, achieving an accuracy of a foot or so requires approximately 48 hours of intermittent access to the signals from the Transit satellites. By contrast, the GPS provides inch-level accuracies with a satellite observation interval lasting, at most, only about one hour.

Determining the Shape of Planet Earth

For thousands of years scientists have tried to determine the size and shape of planet earth. During those centuries, shapes resembling table tops, magnifying glasses, turkey eggs, and Bartlett pears have all, at one time or another, been chosen to model its conjectured shape. The ancient Babylonians, for instance, were convinced that the earth was essentially flat, probably due to erroneous everyday observations. But by 900 B.C., they had changed their minds and decided it was shaped like a convex disk. This belief probably arose when some observant mariner noticed that, whenever a sailboat approaches the horizon, its hull drops out of view while its sail is still clearly visible.

By 1000 B.C. Egyptian scientists had concluded that the earth was a big, round ball. In that era, in fact, Erastothenes managed to make a surprisingly accurate estimate of the actual circumference of the spherical earth. He realized that such an estimate was possible when he happened to notice that at noontime on a particular day, the sun's rays plunged directly down a well at Aswan, but at that same time due north at Alexandria its rays came down at a more shallow angle.

Once he had measured the peak elevation angle of the solar disk at Alexandria on the appropriate day (see Figure 13.1), Erastothenes estimated the distance from Aswan to Alexandria—probably by noting the travel times of sailing boats or camel caravans. He then evaluated a simple ratio to get an estimate for the circumference of planet earth. Translating measurement units across centuries is not an easy thing to do, but our best guess indicates that his estimate for the earth's radius was too large by around 15 percent.[1]

---

[1]Twenty-five centuries later, Christopher Columbus *underestimated* the earth's radius by 25 percent. He wanted to believe that he inhabited a smaller planet so that the Orient would not be prohibitively far away from Europe, sailing west.

THE SCIENCE OF GEODESY

THE EARTH'S RADIUS IS GIVEN BY:

$$R = \frac{L \times 360\,°}{2\,\pi\,\alpha}$$

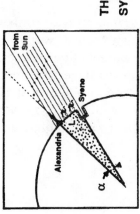

THE DISTANCE BETWEEN ALEXANDRIA AND SYENE (ASWAN) WAS ESTIMATED BY:

- SAILBOAT TRAVEL TIME
- CAMEL CARAVANS (1.2 MI / HR AVERAGE)

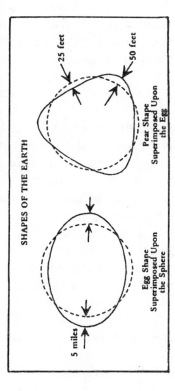

Figure 13.1 In 1000 B.C., the highly insightful Egyptian mathematician, Erastothenes, estimated the radius of the earth by measuring the elevation angle of the sun at Alexandria when it was known to be overhead due south at Aswan (Syene). Then using a simple ratio, he scaled up the measured distance separating those two Egyptian cities to obtain a surprisingly accurate estimate for the circumference of the spherical earth.

179

In 1687, England's intellectual giant, Isaac Newton, displayed his powerful insights when he reasoned that his home planet, earth, must have a slight midriff bulge. Its shape, he reasoned, is governed by hydrostatic equilibrium, as its spinning mass creates enough centrifugal force to sling a big curving girdle of water upward against the pull of gravity. Newton's mathematical calculations showed that this enormous water-girdle must be around 17 miles high. But were the land masses affected in the same way as that bulge of water in the seas? Newton understood that if the earth was rigid enough, the land masses would not be reshaped by centrifugal forces. But he reasoned that, since there were no mountains 17 miles high, the land masses must be similarly affected; otherwise, no islands would peek up through the water in the vicinity of the equator.

Isaac Newton's powerfully intuitive reasoning was entirely correct, but his calculations indicated a flattening factor of 1 part in 230 rather than the 1 part in 300 value that we know, today, accurately characterizes the oblate spheroid of planet earth.

As far as they knew, earthlings occupied a planet shaped like a giant turkey egg until 1958 when an alert researcher, Ann Bailey, at the Goddard Spaceflight Center in Greenbelt, Maryland, carefully examined the reams of tracking data from NASA's first little Vanguard satellite. In a flash, she suddenly realized that the earth must have a small but distinctive pear-shaped component lurking in its gravitational field. The bumps are not actually there, but, because of internal mass concentrations, the earth's gravitational field behaves as though it has three bulging "corners," each about 50 feet high.

In the 1960s, England's Desmond King-Hele and a number of other researchers further refined the shape of our planet into a "wavy" geoid, with surprisingly large numbers of higher-order potential harmonics. More recently, researchers in several different countries have redefined it into a wavy geoid containing lumpy mass concentrations—which twist and distort the orbits of man-made satellites. Today the GPS helps researchers find even better values for the wavy contours of the beautiful blue marble we all call home.

The Theory of Isostasy

Many early geodetic experts were surprised to learn that the force of gravity over the Andes mountains is just about the same as the force of gravity on an airplane at the same height flying over the nearby plains. Why doesn't such a massive mountain chain create extra gravitational forces to tug on any objects huddling nearby? When this curious phenomenon was first observed, some few geophysicists theorized that mountains are light, airy, hollow structures spiking up toward the sky. But actually, the *theory of isostasy* provides a much better explanation. According to the theory of isostasy, a mountain cannot poke its head up for any length of time unless its weight is supported by a big, fat "root" that protrudes down into the denser manma below. Mountains are thus, to some extent, supported by buoyancy forces; they float like timber or icebergs floating on water. Like the oceans, the earth's crust also remains, to a first approximation, in hydrostatic equilibrium.

The Earth's Contours Under Hydrostatic Equilibrium

At many locales, under a variety of conditions, continent-sized regions of the earth, such as Europe, Australia, and North America, have been surveyed and fitted as accurately as possible with oblate spheroids. The resulting maps or "datum-planes" define the equipotential contours of the earth quite well in a small local area.[2] But, at distant locales, the fit may not be particularly snug. For this reason, a few "best fit" spheroids have also been constructed that model the entire earth with reasonable precision. When this global approach is used, the accuracy of the contour in any particular region is sacrificed, but the map coordinates of the spheroid are applicable anywhere in the world, assuming that the local maps have been marked off with the appropriate longitude-latitude coordinates. The WGS-84 (Worldwide Geodetic System 1984), which has been adopted for use as the standard coordinate system by the engineers who masterminded the Navstar Global Positioning System, constitutes an excellent example of a global datum plane.

GPS Calibrations at the Turtmann Test Range

Surveying demonstrations carried out at the Turtmann Test Range in the Swiss Alps have demonstrated that, when a GPS receiver is operated in the carrier-aided (interferometry) mode, it can provide positioning accuracies comparable to those obtained from the finest available laser-ranging techniques. The Turtmann Test Range is nestled in a bowl-shaped valley, thus allowing continuous intervisibility between most of its eight different laser-surveyed benchmarks. The Turtmann Test Range spans 11.6 square miles, with a maximum altitude difference between its eight benchmarks amounting to about 3,000 feet.

Various types of GPS receivers have been tested at Turtmann, most using carrier-aided solution techniques. In particular, the WM101 receiver, jointly produced by Wild Heerbrugg in Switzerland and Magnavox in the United States, has been tested in connection with the POPS surveying software

---

[2]The equipotential contour of the earth is the shape water would assume due to the combined actions of its gravity and the centrifugal force created by its 24-hour rotation.

program to determine how accurately that combination could assemble and process static surveying measurements.

Static Surveying Techniques

Figure 13.2 summarizes the positioning accuracies that the Swiss surveying team were able to achieve in the Turtmann test campaign. In this clever bird's-eye-view depiction of the range, the various baseline lengths are all accurately proportioned. The short vectors are proportional to the surveying errors in the horizontal plane, but they have been magnified 100,000 times, compared with the dimensions of the baseline lengths.

In this particular surveying application the pseudo-range measurements were recorded at each survey site on ordinary cassette tapes. The tapes are later brought together at the base camp and their pseudo-range values were fed into an IBM-PC personal computer. The computer automatically analyzed an overdetermined set of pseudo-range measurements to pinpoint the locations of the pre-surveyed benchmarks and to determine their intervening baseline lengths.

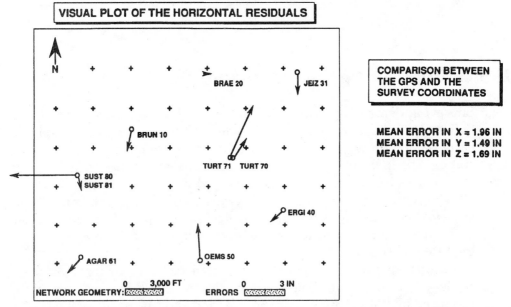

Figure 13.2 The laser-calibrated Turtmann Test Range, located in a bowl-shaped valley in the Swiss Alps, provides surveyors and engineers with an ideal method for testing the accuracy of various GPS surveying techniques. In the test series highlighted in these sketches, the mean positioning errors in the three rectilinear coordinates (x, y, and z) amounted to only 0.2 inches, 0.15 inches, and 0.17 inches, respectively.

In one test series, the 1σ deviations between the GPS measurements and the earlier laser-ranging calibrations turned out to be:

$$\sigma_x = 0.2 \text{ inches}$$
$$\sigma_y = 0.15 \text{ inches}$$
$$\sigma_z = 0.17 \text{ inches}$$

In an earlier test involving only four base stations with three unknown baseline lengths of 382.2 feet, 1,644.4 feet, and 333 feet, the average surveying errors were:

$$\sigma_x = 0.2 \text{ inches}$$
$$\sigma_y = 0.35 \text{ inches}$$
$$\sigma_z = 0.35 \text{ inches}$$

Both sets of measurements were estimated using static surveying techniques in which the GPS receiver sits at each site for about a half hour to record several hundred pseudo-range measurements. All of the measurements from the various sites are then processed simultaneously to achieve the desired results.

Kinematic and Pseudo-kinematic Surveying

Static positioning techniques of the type used in Switzerland's Turtmann Test Range yield surprisingly accurate positioning estimates, but obtaining the required observations is time consuming and labor-intensive.

An alternate approach called *kinematic surveying* speeds the observation processes and reduces labor costs. In kinematic surveying the GPS receiver is carried around a closed loop connecting the various survey sites. The closed loop traverse helps eliminate some of the solution ambiguities routinely encountered in conventional static surveying techniques.

When kinematic surveying is employed, site occupancy times are reduced from a half hour or so to only one or two minutes. Substantial productivity gains result, but a continuous-loop traverse is required and the receiver must maintain an unbroken lock on the carrier waves of the same four satellites throughout the continuous traverse. At the completion of the traverse, an "antenna swap" is performed between the stationary receiver and the portable receiver that has been carried around the continuous loop. In an antenna swap, the two antennas are physically interchanged. This helps the surveyors detect any cycle slips that may have occurred during the measurement procedures.

Another approach called *pseudo-kinematic surveying* is also saving time and labor. In a pseudo-kinematic navigation solution, two brief sets of pseudo-range measurements are recorded at each survey site an hour or so apart. This allows the satellite viewing geometry to change as the GPS satellites sweep across the sky. Pseudo-kinematic surveying enhances the productivity and efficiency of professional surveying teams while producing accuracies roughly comparable to conventional kinematic surveying. Field test results indicate that the optimum amount of time between observations typically ranges from 50 to 105 minutes.

Recent tests conducted in Arizona's Superstitious Mountains have demonstrated conclusively that pseudo-kinematic surveying can be a highly accurate, cost-effective surveying technique. Substantial labor savings definitely result, but the pseudo-kinematic surveying also reduces the amount of computer time needed to process the data. Moreover, pseudo-kinematic surveying uses essentially the same software as static GPS surveying, and it eliminates the need to maintain continuous code-lock and carrier-lock tracking throughout the observation interval.

Figure 13.3 summarizes the salient features of four different types of GPS positioning-fixing techniques. The first one—differential navigation—can be used in dynamic operating environments. The last three—static, kinematic, and pseudo-kinematic surveying—all employ carrier-aided interferometry solutions to fix the locations of benchmarks with extreme accuracy. Static surveying is still quite popular among professional surveyors. But, under favorable conditions, the kinematic and pseudo-kinematic approaches can save time and money in large amounts for professional surveyors who earn their livings in a highly competitive industrial environment.

Freeway Surveying During War in the Persian Gulf

During the Persian Gulf war, Southern California surveying operations were affected quite unexpectedly, by military events half a world away. Shortly after Iraqi troops pushed their way into Kuwait, Air Force technicians began reconfiguring the Navstar constellation to better serve the troops engaged in Operation Desert Shield. As the satellites were gradually moved into new positions, the number available to surveyors in the Los Angeles area began to decline. At that time surveying teams from the Psomas Company (Costa Mesa, California) were using the satellite signals to survey and remap the Riverside Freeway. Unfortunately, their five-satellite coverage intervals shrank from 10.5 hours a day to only 3.5 hours a day, a coverage loss that was very hard to make up. Once the 24-satellite constellation is complete, of course, coverage will be provided everywhere 24 hours every day.

Navstar Positioning for Landsat D

The Landsat D, an earth-resources satellite, was one of the first orbiting space vehicles rigged to fix its position using the signals being broadcast by the

SURVEYING TECHNIQUES USING THE GPS

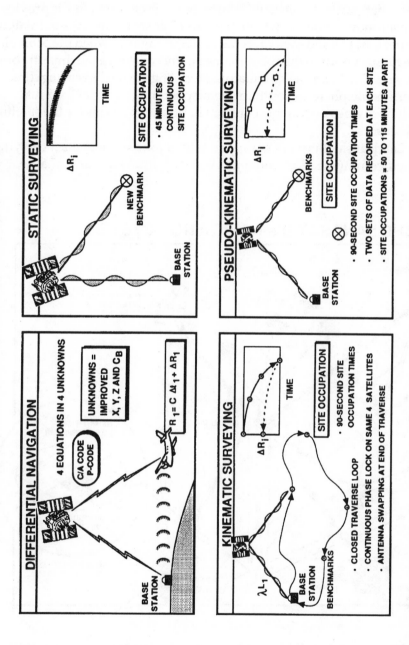

Figure 13.3 Capsule descriptions for four different GPS navigation methods are highlighted in this figure. The last three, all of which employ carrier-aided position-fixing techniques, are used by surveyors in low-dynamic environments to determine benchmark positions to within a few inches, or even less. Static surveying was the first approach that was widely used. The last two, kinematic and pseudo-kinematic surveying, were developed to save time, energy, and money while achieving roughly comparable surveying results.

Navstar constellation. As Figure 13.4 shows, the Landsat D picks up the signals from the Navstar satellites to pinpoint its position so that it can make precisely calibrated false-color images of the earth below. The Landsat D was launched into a 380-nautical mile orbit, with an inclination of 98.2 degrees. That particular orbital location was selected because it provides a sun-synchronous orbit in which the orbit plane twists approximately 1 degree per day. In a sun-synchronous orbit the angle between the orbit plane and the radius vector to the sun remains nearly constant at all times through-out the year. The Landsat D's equatorial crossings in the *descending* direction (northern hemisphere to southern hemisphere) all occur at 9:45 A.M. local time. This provides reasonably constant lighting conditions for the satellite's optical sensors.

The Landsat satellites are equipped with a long vertical mast carrying a high-gain communication antenna. A stubby little GPS antenna near the

THE LANDSAT D's NAVIGATION SYSTEM

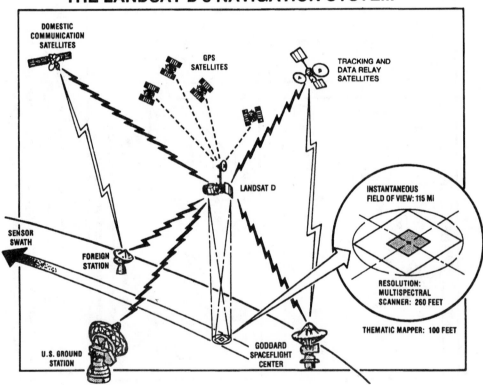

Figure 13.4 Position-fixing services for the Landsat D earth-resources satellite were pro-vided by the GPS satellites that orbited the earth at much higher altitudes. When the full GPS constellation has been installed in space, Landsat's average positioning errors are expected to amount to only about 20 feet. When the GPS was first used for this purpose, the navigation signals from the rather limited constellation were frequently blocked by the earth, so errors of several hundred feet quickly began to accumulate.

center of that mast picks up the signals from the Navstar satellites so that the receiver inside can accurately determine its position in space.

The Landsat's Spaceborne Receiver

The Landsat D's spaceborne receiver, which is about as big as a breadbox, was designed and built by Magnavox engineers at their facility in Torrance, California. It is a two-channel parallel receiver capable of handling the moderate dynamic uncertainties compatible with the motion of an orbiting satellite. The Magnavox receiver weighs 43 pounds and draws 45 watts of electrical power when it is performing navigation functions. In its standby mode—when the GPS satellites are all situated below the horizon—it draws only about 10 watts.

On-Orbit Navigation Accuracy

Computer simulations indicate that, with the full 24-satellite GPS constellation in place, the Magnavox spaceborne receiver should yield an average positioning error of only about 20 feet (CEP). Surprisingly, a low-altitude satellite can establish its position more accurately than a comparable user located on the ground. This stems from the fact that the L-band navigation signals do not pass through the earth's ionosphere and the troposphere and because the no mask angle need be imposed. Moreover, an orbiting satellite travels around its orbit in such a regular and predictable manner, high dynamic uncertainties are seldom encountered.

The Landsat D earth-resources satellite was launched into orbit when only a few Navstar satellites were available to help fix its position. Consequently, its initial navigation accuracies were not particularly impressive. During those intervals when Landsat's navigation antenna was illuminated by signals from four or more GPS satellites, it provided excellent accuracy. But, as the GPS satellites began to pass below the horizon, its accuracy quickly degraded. When coverage was skimpy, the Landsat's onboard computer had to use astrodynamics calculations to estimate its current position. Fortunately, less than 90 minutes later, when Landsat again caught up with the small constellation of Navstar satellites, its navigational error was again reduced to a much smaller value.

Orbit Determination for High-altitude Satellites

The full GPS constellation can provide accurate navigation solutions for low-altitude and medium-altitude satellites, but, when a satellite is orbiting above the 10,898–nautical mile GPS altitude, the earth blocks the main beam, thus keeping the strongest L-band signals from illuminating its GPS antennas. GPS navigation is still possible, but only if the high-altitude satellite can pick up the weaker side portions of its main beam or its even weaker side lobes. With an excellent high-gain antenna, enough signal strength is available for marginal navigation coverage, but the viewing geometry is substantially degraded compared with the geometry available to low-altitude satellites.

Navigational errors for the low-altitude satellites are sometimes as small as 15 feet. For intermediate altitudes (below 10,000 nautical miles) the errors are more like 65 to 100 feet, and, for satellites at the geosynchronous altitude, the errors range as high as 300 feet or slightly higher. However, even at the geosynchronous altitude, GPS positioning is still considerably more accurate than the tracking services provided by the Satellite Control Facility operated by the DoD, which employs various types of optical and radar devices to skin-track a large number of orbiting satellites. For low-altitude and medium-altitude satellites, the Satellite Control Facility achieves navigational errors in the 300-foot range. Accuracies decline for satellites at higher altitudes, because the baselines available to the ground-based Satellite Control Facility do not exceed the diameter of the earth. At geosynchronous altitudes and beyond, the angular extent of the baseline shrinks so much that the positioning errors turn out to be about 800 feet.

Of course, any accuracy comparisons between the Satellite Control Facility and the GPS are intrinsically unfair because GPS tracking is available only for satellites that carry a GPS receiver onboard. DoD tracking is available for any orbiting object large enough to track with radar or optical sensors, whether it cooperates or not.

Today's Available Spaceborne Receivers

As Table 13.1 indicates, a number of user-set makers (Magnavox, Texas Instruments, Interstate Electronics, Motorola, Rockwell International) have all produced—or are on the verge of producing—Navstar receivers designed for use aboard satellites in space. The engineers at Magnavox, for example, developed the special P-code receivers carried onboard the Landsat earth resources satellites. Magnavox has also sold other units to the U.S. Air Force for unspecified military purposes.

Texas Instruments in Lewisville, Texas, has provided at least two 2-channel multiplexing P-code receivers to military customers, the first of which was launched as early as 1981. In addition, Texas Instruments' engineers are working on a contract with the U.S. Air Force, with possibilities for other shuttle-based and satellite-based units.

Interstate Electronics in Anaheim, California, has flown several spaceborne translators aboard the Trident and other strategic missiles. A translator does not yield real-time positioning solutions. It merely rebroadcasts the raw L_1 and L_2 navigation signals to ground-based processing units, which later execute the desired navigation solutions. Interstate Electronics has devel-

Table 13.1 GPS spaceborne receiver candidates

Vendor	Receiver description	Weight (pounds)	Power (watts)	Accuracy (SEP)	Space flight experience	Comments
Magnavox Torrance, California	GPS-PAC 2-channel P-code Receiver	43	45 (10 in standby mode)	< 10 meters (full constellation assumed)	2 units flown aboard landsats 4 and 5 Other units sold to the Air Force	Flight unit was ground receiver modified for space use It suffered from logic upsets due to cosmic rays
Texas Instruments Plano, Texas	2-channel multiplexing P-code receiver	100	40	1 meter pseudorange accuracy	More than one flown into space Space flights began in 1981	Excellent radiation hardening Dual redundant design with crosstrapping
Texas Instruments Plano, Texas	PSN-9 5-channel C/A-code receiver	14	16	60 meters	Ground-based version of receiver under Contract to the JPO Possibilities for shuttle launch	Spaceborne receiver would be based on TI 420 Radiation-hardened
Interstate Electronics Anaheim, California	Multi-channel P-code translator	3.2	Not available	A few meters	Several have flown into space aboard Trident missiles Position solution on the ground	Translator approach requires ground station along the line of sight

(*Continued*)

Table 13.1 (*Continued*)

Vendor	Receiver description	Weight (pounds)	Power (watts)	Accuracy (SEP)	Space flight experience	Comments
Motorola Scottsdale, Arizona	12-channel P-code receiver interferometry mode	9	35	2 to 6 meters pseudo-range 0.2 to 0.6 meters carrier-aided	2 interferometry mode receivers under contract to JPL for the Topex spacecraft	Interferometry mode receivers highly accurate but easy to jam
Rockwell International Anaheim, California	2-channel P-code receiver	9	8	~ 8 meters	4 units plus 2 options under contract to MSFC for the OMV (orbital maneuvering vehicle)	
Rockwell International Anaheim, California	2-channel P-code Receiver	1	5	16 meters	Working toward lightsat applications	Receiver to be based on the chipset used in the Rockwell Collins "Virginia Slim" receiver

oped multichannel translators weighing as little as 3.2 pounds. Even smaller units are in the works.

Motorola of Scottsdale, Arizona, has developed a 12-channel P-code receiver named after the Monarch butterfly. Motorola's Monarch performs highly accurate carrier-aided interferometry solutions for orbiting satellites. Its pseudo-ranging solutions are accurate to within 20 feet CEP, and its carrier-aided solutions typically yield positioning solutions accurate to within 1 or 2 feet.

Rockwell International in Anaheim, California, was under contract to develop four units (plus two options) for use aboard NASA's Orbital Maneuvering Vehicle. Research efforts progressed to a sophisticated state of readiness for producing the desired receivers, which were to weigh 9 pounds and consume only 8 watts of electrical power. Unfortunately, budget cuts forced the cancellation of these spaceborne receivers. Later, the entire Orbital Maneuvering Vehicle itself was canceled.

Rockwell International is also working on a 2-channel P-code unit using the solid-state chip-set from their hand-held *Virginia Slim receiver*. This "mighty mite," which is to weigh only about 1 pound, is being designed for use aboard lightsats—small, inexpensive, limited-purpose satellites widely touted for their compact packaging and impressive performance results.

Now that the capabilities of the Navstar system have been demonstrated so convincingly here on earth, satellite makers all around the world are even more anxious to harness its power and versatility for a broad range of missions in space. When they do, whole new marketing arenas will doubtlessly emerge, with an impressive new array of bells and whistles.

14

Military Applications

In early bombing raids against the Nazi war machine, only about 3 percent of British bombs landed within 5 miles of their intended targets. But, during the Persian Gulf War, precision-guided munitions consistently struck much smaller targets with pinpoint accuracy thanks, in part, to the precise positioning services provided by a partial constellation of Navstar navigation satellites.

In one widely publicized incident, signals from the Navstar satellites directed two SLAMS (Standoff Land-Attack Missiles) toward the walls of an Iraqi power plant. The first missile blasted a hole in the wall. Two minutes later, the second flew through the hole to blow up the electrical equipment inside.

When studies revealed that Iraqi troops had managed to obtain, at most, only about two dozen Navstar receivers, American military commanders decided to eliminate selective availability, thus giving even simple civilian receivers full military precision. Every available Navstar receiver—military and civilian—was then shipped to coalition forces in the Persian Gulf. Trimble Navigation sent 1,000 SLGR's (Small Light-Weight Receivers), or "sluggers," as the troops affectionately called them. Magellan shipped even larger numbers, and other manufacturers provided as many as they could manage to produce.

Modern electronic technology, supplemented by old-fashioned Yankee ingenuity, made an unbeatable combination along the Persian Gulf. Iraqi troops tried to maintain their courage long enough to participate in "The Mother of All Battles," so enthusiastically touted by Saddam Hussein. But, by the time the ground war finally came, they were so dazed and demoralized by America's technological might, their will to resist had ebbed away.

"My unit fired only once during the war," said one Iraqi artillery officer who was still amazed by the devastating response. "A few minutes later, one-third of my troops ran away. The other two-thirds were dead."

Soon, white flags were fluttering over the wind-swept sands of Iraq and Kuwait and, when thousands of Iraqi stragglers were rounded up, many knelt in gratitude to kiss the hands of their captors. One ragged Iraqu unit was so intent on capitulation, its members actually tried to surrender to an unmanned remotely piloted drone!

The Military Benefits of the Worldwide Common Grid

One of the hidden military benefits of Navstar navigation is that it provides a worldwide common grid for coordinating diverse military activities. WGS-84 (Worldwide Geodetic System 1984) has been adopted as the standard GPS coordinate system. But military receivers can be operated with any of 45 popular datum planes (map coordinate systems) by selecting from the menu items on a user-friendly display. In military confrontations the use of a single consistent set of map coordinates greatly reduces the complexity of coordinating military operations that, in the past, were often handled using a patchwork quilt of different datum planes. If two military units are inadvertently operating in incompatible datum planes, mismatches of several hundred feet can be encountered without anyone realizing that an error has occurred. With GPS navigation, serious mismatches are still possible, but they are much less likely.

Field Test Results

Performance results obtained in four different types of military tests—static positioning, marine navigation, aerial rendezvous, and bomb delivery—are summarized in Figure 14.1. Notice that, even in early *static positioning* tests, root-mean-square errors as small as 23 feet were routinely achieved. In recent years these results have been further improved, especially when military researchers have used more sophisticated positioning techniques, such as differential navigation and carrier-aided solutions.

Marine navigation was tested aboard the USS Fanning as it sailed out of Acapulco Harbor. Throughout the test, Navy navigators compared GPS positioning solutions with sextants sightings, optical measurements, and other shipboard navigation techniques.

In the *aerial rendezvous* sequences, an F-4 aircraft rendezvoused repeatedly with a C-141 tanker. In six independent rendezvous tests the maximum separation error turned out to be approximately 100 feet. In other words, the worst-case miss distance was only slightly larger than the wingspan of the C-141.

Bomb delivery tests using six Mark 82 low-drag dumb bombs in six straight-and-level passes at 10,000-foot altitude produced a maximum dis-

GPS FIELD TEST RESULTS

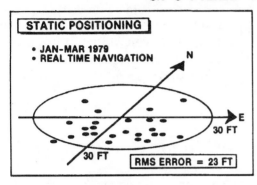

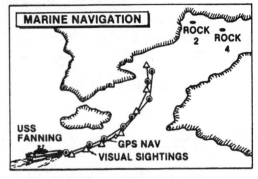

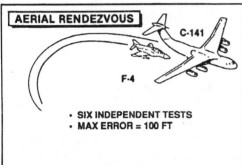

Figure 14.1 Early military experiments using the Navstar Global Positioning System showed that its root-mean-square error for static P-code navigation amounted to only about 23 feet. The Navstar system was also tested in connection with marine navigation when the USS Fanning sailed out of Acapulco Harbor and for aerial rendezvous between an F-4 aircraft and a C-141 tanker with equally impressive results. Bomb delivery missions involving six straight-and-level passes at 10,000 feet with MK-82 low-drag bombs produced a maximum dispersion error (worst-case impact) of only 56 feet.

persion error of approximately 56 feet. Figure 14.1 shows how the impact points look when they were superimposed on a large, metal bridge. Actually, no bridge was present during these bomb-delivery tests, which took place over the desert at Yuma, Arizona. The artist sketched the bridge in later, but she did position and scale it as accurately as she could.

Projected Battlefield Benefits

The military test results summarized in Figure 14.1, coupled with dozens of other similar tests, were fed into computer simulation programs to determine how GPS-equipped military forces are likely to perform in actual combat situations. The four sketches in Figure 14.2 highlight a few of the results obtained from these computer simulations, most of which compared the effectiveness of GPS-equipped military forces with the same forces using more conventional navigation systems, such as Loran, Omega, and inertial

PROJECTED BATTLEFIELD BENEFITS

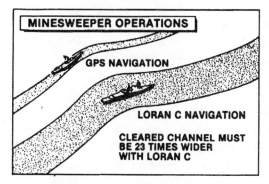

MINESWEEPER OPERATIONS

GPS NAVIGATION

LORAN C NAVIGATION

CLEARED CHANNEL MUST
BE 23 TIMES WIDER
WITH LORAN C

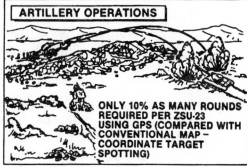

ARTILLERY OPERATIONS

ONLY 10% AS MANY ROUNDS
REQUIRED PER ZSU-23
USING GPS (COMPARED WITH
CONVENTIONAL MAP –
COORDINATE TARGET
SPOTTING)

PRECISION TARGET INTERDICTION

TARGET KILLS PER SORTIE
AGAINST TANKS, SAM SITES,
AMMO DUMPS, AND RADAR
INSTALLATIONS INCREASED
BY 400 TO 600%

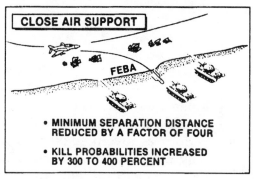

CLOSE AIR SUPPORT

FEBA

• MINIMUM SEPARATION DISTANCE
REDUCED BY A FACTOR OF FOUR

• KILL PROBABILITIES INCREASED
BY 300 TO 400 PERCENT

Figure 14.2 Force-multiplier effects achieved when the GPS is used in connection with various types of military missions range from 300 percent for close air support to 2,300 percent for the Navy's minesweeper operations. Two separate computer simulations were executed for each mission. In the first case the mission was conducted without GPS guidance. In the second case GPS guidance helped enhance mission effectiveness.

navigation. These broad-ranging simulations yielded the following specific conclusions:

1. On the average, minesweeping operations are about 20 times more effective when forces equipped with GPS receivers are compared with forces using more conventional radionavigation techniques. These simulation results assume that both the minesweepers and the ships subsequently traversing the swept channel are rigged with GPS receivers.

2. GPS-equipped artillery batteries attempting to destroy a Soviet ZSU-23 radar-controlled anti-aircraft gun typically require only about 10 percent as many rounds to achieve a successful kill, compared with those who are attacking a similar target using optical observations and conventional map-coordinate target-spotting techniques.

3. Precision target interdiction missions against SAM sites, ammunition dumps, and radar installations can destroy 400 to 600 percent more enemy hardware if the attacking aircraft employs ordinary "dumb"

bombs delivered by means of GPS-based precision bombing techniques.

4. Close air support missions can produce kill-probability improvements, ranging from 300 to 400 percent if signals from the GPS satellites are used to determine the aircraft's position and velocity prior to bomb release.

Other tests, computer simulations, and think-tank studies point toward comparable force-multiplier effects for a much wider variety of military missions. These include strategic bombing, air-to-air intercept, air-mobile ICBM delivery, base station positioning, and military search-and-rescue operations.

Test Range Applications

Military test ranges are used in evaluating ordnance devices, such as air-to-air missiles, artillery projectiles, and low-drag bombs, to make sure they will perform as advertised during hostile actions against enemy forces. Historically, most military test ranges have employed ground-based radionavigation systems to track and position the ordnance devices during carefully controlled reenactments of realistic battlefield encounters. But independent studies conducted by The Analytical Sciences Corporation have shown that using GPS receivers to pick up signals from the Navstar satellites could save both time and money for test range applications. The cost components for operating eight selected military test ranges equipped with ground-based navigation aids were compared with the cost components that would likely be incurred if GPS positioning and tracking devices were used instead. Over a 20-year interval the amount of money that could be saved by switching to GPS equipment at eight representative test ranges turned out to be $380 million. When researchers at the Analytical Sciences Corporation extrapolated this result to include all of the American military test ranges in operation, the 20-year saving was projected to be between $800 million and $1.2 billion.

For many test range applications, large numbers of high-speed vehicles and their ordnance devices must be tracked continuously to verify the desired test result. In most cases, existing ground-based systems provide acceptable accuracy, but the tracking equipment suffers from a variety of fairly serious shortcomings. In particular, the navigation signals broadcast by ground-based transmitters are often shielded by mountain peaks, hills, and trees, and they can be jammed or saturated by stray electromagnetic signals. The proliferation of nonstandard and incompatible ground-based devices at Military Test Ranges also causes difficulties with data coordination, engineering analysis efforts, and hardware modifications and maintenance. Among other advantages, the emergence of the GPS system provides a genuine opportunity for standardization and simplification of military test range hardware with similar simplifications in experimental protocols and procedures.

Translator-type receivers are advantageous for many different types of test range applications. A translator does not perform a navigation solution to obtain its position, velocity, and time. Instead, it merely picks up the L_1 and/or L_2 signals being broadcast by four or more GPS satellites, and rebroadcasts their raw, unprocessed L-band signals to ground-based receivers on a different frequency, typically S-band. When the S-band downlink reaches the ground, it is recorded on wideband magnetic tapes and, in some cases, computer-processed in real time to establish the instantaneous position and velocity of the translator. If necessary, this time-varying state vector can be relayed back up to the moving platform on another S-band transmission link. In any case, precise post-processing procedures can be used to reconstruct extremely accurate ordnance trajectories after the military test is complete.

For some experiments, translators have been rigged to work with P-code navigation signals, but engineering studies indicate that acceptable accuracy can be achieved for 90 percent of all test range applications by employing the simpler C/A-code signals. This selection also allows the S-band relay to use a much smaller bandwidth (2 megahertz instead of 20 megahertz).

Researchers at the Analytical Sciences Corporation have assembled a collection of accuracy estimates for applications using different GPS navigation techniques for test range applications. Throughout their study they assumed a conservative PDOP (Position Dilution of Precision) value of 3. They concluded, as Table 14.1 indicates, that real-time, absolute C/A-code

Table 14.1 Expected GPS Accuracy levels for range instrumentation applications

ASSUMPTIONS: NO UNCOMPENSATED USER DYNAMICS
(INS-AIDED)
RECEIVER OR TRANSLATOR
GDOP = 3
1 σ ERRORS

	Real Time*		Real-time Differential		Post-mission Differential	
	C/A-Code	P-Code	C/A-Code	P-Code	C/A-Code	P-Code
Position (ft) x, y	30	14	25	7	6	2
z	51	23	41	12	10	4
Velocity (fps)† x, y	0.06–0.6	0.06–0.6	0.06–0.6	0.06–0.6	0.02	0.02
z	0.1–1.0	0.1–1.0	0.1–1.0	0.1–1.0	0.03	0.03

*Assumes 10 ft uncompensated ionospheric bias, C/A not available on L_2, and 10 ft multipath error.
†for 1–10 Hz data rates.

navigation would provide a 1σ positioning error in the horizontal plane of approximately 30 feet. The comparable real-time P-code navigation error was 14 feet in the horizontal plane. Smaller errors can be obtained by using real-time differential navigation, which results in a P-code error of 7 feet, and P-code post-mission differential, which results in an error of only 2 feet.

Bomb delivery missions using the GPS navigation signals have yielded surprisingly small dispersions. In a bombing mission, the GPS receiver is usually integrated with various other avionics systems carried onboard the aircraft. Working together, the sensor readings from all the avionics devices compute the "instantaneous impact point" of the bomb being dropped. The instantaneous impact point calculations take into account the effects of gravity, drag, wind, coriolis force, ejector foot-force delays, and so on, in determining where the bomb would land if it was dropped at successive instants along the flight. When the instantaneous impact point crosses the specified target, a bomb is automatically released.

Flight tests using a variety of bomb delivery modes have been conducted, mostly over the desert at Yuma, Arizona. These tests have included, among others, toss bombing, dive bombing, and level-laydown carpet bombing techniques. Hundreds of bombs have slammed into the desert floor at Yuma, which now resembles a pockmarked lunar landscape. Generally speaking, the impact dispersions have been similar to the results reviewed in Figure 14.1.

Military Receivers

The Collins Government Avionics Division of Rockwell International produced the first large batch of military receivers using a highly automated production line at Cedar Rapids, Iowa. However, in a more recent competition, SCI of Huntsville, Alabama, displaced Rockwell as the DoD's main supplier for the standard military GPS receivers. According to *Aviation Week and Space Technology*, SCI's bid to produce 6,000 military receivers totaled $175 million, compared with the Rockwell bid of $220 million dollars. Other contractors, including Interstate Electronics, Trimble Navigation, Magnavox, and Rockwell International, are producing various types of military receivers under a variety of other DoD contracts.

Carrier-landing Accuracies

As Figure 14.3 shows, rather dramatically, the GPS is considerably more accurate than most competing navigation systems. In this sketch the aircraft carrier, USS Nimitz, is drawn to the same scale as the six ovals, each of which represents the error ellipse for one of six different types of navigation systems. Notice that Omega navigation involves a relatively large error of 7,200 feet, as represented by an error ellipse several times bigger than the

NAVIGATION ACCURATE COMPARISONS

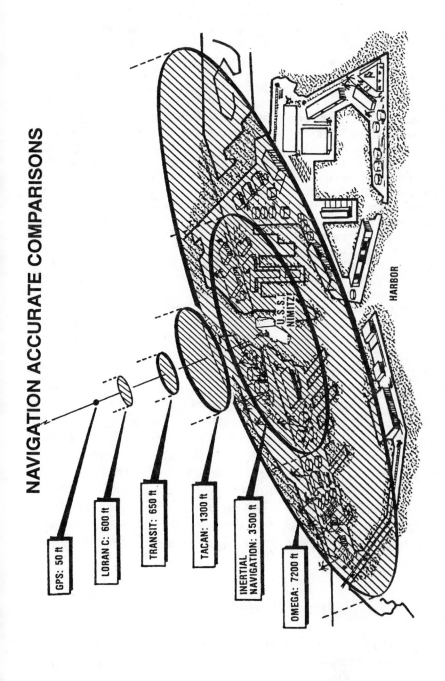

GPS: 50 ft

LORAN C: 600 ft

TRANSIT: 650 ft

TACAN: 1300 ft

INERTIAL NAVIGATION: 3500 ft

OMEGA: 7200 ft

U.S.S. NIMITZ

HARBOR

Figure 14.3 The diameters of the increasingly smaller ovals in this figure are proportional to the positioning errors associated with six different types of navigation systems that might be used by a pilot in preparing to land his plane on the deck of an aircraft carrier. If you were a carrier pilot, which type of landing aid would you want the technicians to install in your plane before you tried to set it down on the carrier's deck?

Routing Ships on the High Seas

Researchers and technicians at Oceanroutes in Palo Alto, California, earn their daily bread using three different types of satellites for finding safe and efficient trajectories for large oceangoing vessels. Each optimum route takes into account real-time weather conditions, the physical characteristics of the ship, and the wishes of the ship's master—who is given an updated trajectory twice each day. The Navstar constellation provides accurate positioning information that is relayed from the ship to Palo Alto through INMARSAT satellites. Weather satellites from various countries furnish the necessary meteorological reports. Sitting in their comfortable offices in Palo Alto and in several other cities around the globe, Oceanroute's engineers work with more than a thousand ships in a routine month. Each recommended route is custom designed for that particular ship "on that specific voyage, with the given cargo load, status of trim and draft, with the ship's own distinctive speed and sea-handling characteristics."

The computer program emphasizes emerging weather, but is also takes into account currents, fog, choke points, navigational hazards, and sea ice in northern regions. Some cargoes, such as fruit and oil, are temperature-sensitive; others, such as automobiles and heavy machinery, may shift under heavy waves. Still others have time-critical deliveries. The Oceanroute's program successfully takes these and numerous other factors into account whenever it makes its routing recommendations.

The cost of the service for a typical voyage is $800, a fee that is repaid 30 to 40 times over by shortened travel times and more efficient maritime operations. In 43,000 crossings aided by Oceanroute's computers, travel times have been reduced an average of four hours in the Atlantic and eight hours in the Pacific. Operating a large oceangoing vessel can cost as much as $1,000 per hour, so time savings alone can translate into enormous reductions in cost. Other expenses are also reduced. When Oceanroute's services were not yet available, the cost of repairing weather-damaged ships ran from $32,000 to $53,000 in an average year. Today, for some companies, these costs have plummeted to only about $6,000. Cargo damage has also declined. One international auto dealer told a team of Oceanroute's researchers that his cargo damage claims had dropped by over $500,000 per year.

aircraft carrier. An inertial navigation system that has been drifting, without updates, for approximately one hour operates with an average error about half as large: 3,500 feet. The Tacan and the Transit Navigation System involve 1,300-foot and 650-foot navigation errors, respectively, under average conditions, whereas the navigation error for Loran C is about 600 feet. These error magnitudes are compared with the GPS error of only 50 feet, as represented by the little dot at the top of the figure. Landing on the deck of an aircraft carrier using the GPS is thus considerably easier and safer than using any of the other popular navigation systems depicted in Figure 14.3.

Amphibious Warfare Operations

Close air support for amphibious beach landings has also been carefully analyzed for military missions with and without GPS positioning (see Figure 14.4). In each of the two cases considered, 182 low-drag dumb bombs were dropped from 30,000 feet into the two rectangular target zones loaded with enemy artillery and tanks. With the relatively inaccurate navigation provided by the Loran C, the bombardiers could safely carpet bomb only about

CLOSE-AIR SUPPORT OF AMPHIBIOUS LANDINGS

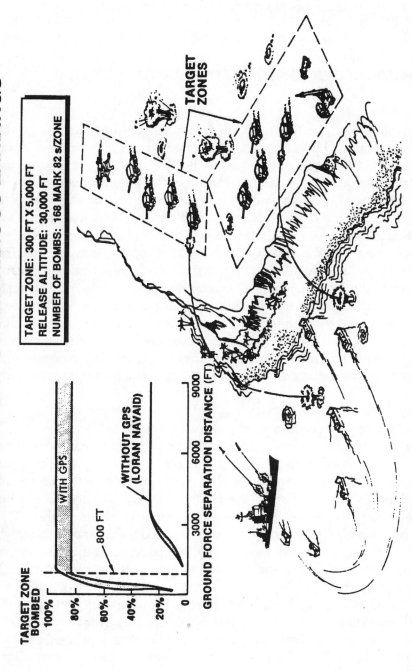

Figure 14.4 The enemy artillery emplacements in this figure are to be carpet bombed by B-52s in support of an amphibious beach landing. If the planes are guided by GPS receivers, they can bomb 80 to 90 percent of the two target zones without substantial risk of hitting their own troops. By contrast, if they are guided by less accurate Loran navaids, only about 20 percent of the target zones can be safely bombed.

20 percent of the two target zones; otherwise, they risk hitting their own troops. But, with the more accurate navigation provided by the GPS, they could safely bomb 80 to 90 percent of the two target zones, thus substantially enhancing the number of enemy installations they could destroy.

Accuracy-enhancements for Strategic and Cruise Missiles

In times of armed conflict, the GPS is not designed for use in guiding intercontinental ballistic missiles and other strategic weapons toward enemy targets thousands of miles away. But GPS navigation is being used to improve the delivery accuracy of some of the missiles operated by the DoD. The Trident *accuracy improvement program* provides an instructive example. Trident is a three-stage solid-fueled rocket launched from a submerged submarine. During its test flights the missile is tracked along its trajectory using the GPS signals together with supplementary signals from one or more pseudo-satellites. GPS translators relay the tracking information down to the ground, where it is computer-processed to reconstruct the trajectory of the Trident missile with special emphasis on its warhead's splashdown point. By monitoring a series of missions of this type, military planners can find and correct the causes of impact errors to refine the mission performance of the Trident's inertial guidance system. This process continually improves impact accuracy, to give the missile hard target kill capabilities.

None of the 200 basing modes that DoD researchers proposed and evaluated for the MX "Peacekeeper" missile called for the onboard use of a GPS receiver. But the airborne basing mode, in which the MX missile was, in time of war, to be released from an aircraft in flight, could have benefitted from the precise position and velocity measurements provided by a GPS receiver onboard the aircraft. Early studies indicated that targeting accuracies could be substantially improved in this manner without relying on the GPS to guide the missile's flight. However, even with GPS guidance onboard the aircraft, the airborne delivery mode would still be considerably less accurate (but more survivable) than the silo basing mode ultimately embraced by Pentagon planners.

The Tomahawk Cruise Missile knocked out several difficult targets during the Persian Gulf War. Tomahawk flies subsonically over a range of 1,500 miles along a winding, defensive trajectory skimming over the treetops as low as 100 feet (see Figure 14.5). Its inertial navigation system is periodically reinitialized using terrain-matching techniques in which optical sensors on board the missile match the contours on the ground below against electronic contour maps previously stored in the memory of its computer.

The Tomahawk Cruise Missiles used in the Persian Gulf were not guided to their targets using the GPS. But a Tomahawk upgrade program calls for

GPS FOR TOMAHAWK UPGRADE

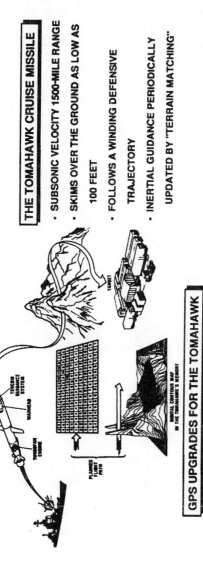

THE TOMAHAWK CRUISE MISSILE

- SUBSONIC VELOCITY 1500-MILE RANGE
- SKIMS OVER THE GROUND AS LOW AS 100 FEET
- FOLLOWS A WINDING DEFENSIVE TRAJECTORY
- INERTIAL GUIDANCE PERIODICALLY UPDATED BY "TERRAIN MATCHING"

PLANNED FLIGHT PATH

TERCOM GUIDANCE SYSTEM

WARHEAD

TURBOFAN ENGINE

TARGET

DIGITAL CONTOUR MAP IN THE TOMAHAWK'S MEMORY

GPS UPGRADES FOR THE TOMAHAWK

- 2-CHANNEL NAVSTAR RECEIVER
- ADVANTAGES OF GPS INSTALLATION:
  - IMPROVED TARGETING
  - REDUCED MISSION PLANNING TIME
  - ENHANCED NAVIGATION ACCURACY

|← 8 INCHES →|

SOURCE: "GPS FOR TOMAHAWK UPGRADE", THE GPS QUARTERLY. (ROCKWELL COLLINS) VOL. 4, NO. 2, JULY, 1989.

Figure 14.5 The Tomahawk cruise missiles fired during the Persian Gulf War were guided to their targets by inertial navigation systems updated periodically by terrain-matching techniques. Future Tomahawks will be rigged to carry 2-channel GPS receivers. The GPS navigation aids will improve targeting, reduce mission planning time, and enhance the navigation accuracy for a broad range of important long-range Tomahawk missions.

the installation of a 2-channel GPS receiver that will help improve targeting, reduce mission planning time, and enhance navigation accuracy for future missions.

In addition to the military missions specifically described in the preceding paragraphs, Navstar navigation is yielding important benefits to a number of other military programs, including the rapid deployment force, Minuteman accuracy improvement, over-the-horizon radar-sensing techniques, and military search-and-rescue operations. As the Persian Gulf conflict amply demonstrated, the Navstar GPS has exceptional capabilities for providing surprisingly large force-multiplier effects for a broad range of important military missions.

15

Civil Applications

The Navstar Global Positioning System is financed by the DoD, but it is being used by a surprisingly broad range of nonmilitary users. As Figure 15.1 indicates, many of its applications have been ending up on the nightly news. The pilots of the *Voyager* aircraft, for instance, used Navstar navigation to help guide their sleek little craft on its nonstop flight around the world. When they finally landed, both pilots agreed that, if the Navstar signals had been available throughout the entire journey, they would have had a much easier time completing their long and grueling mission.

When two chubby little robots managed to locate the wreckage of the *Titanic* in the North Atlantic, signals from the Navstar satellites were used to mark the spot so that others could later return. The crews that dug a tunnel under the English Channel linking Dover with Calais were proud and happy that GPS position-fixing proved to be so beneficial. The project engineers made sure that the two tunnels would meet in the middle by using GPS receivers to position and align laser beams at the two ends of the tunnel for use in guiding the French and British tunneling crews.

Navstar receivers can also aid long-wall coal mining in novel and interesting ways. Underground coal miners usually leave about 20 percent of the coal in place as massive pillars to support the roof. But in a long-wall coal mine, a big, underground machine slices away the coal along a linear seam, taking practically all of it away. The operator is protected from falling debris by a heavy slanting metal roof.

Long-wall coal mining is an efficient way to produce coal, but gradual subsidence of the soil above tends to damage expensive structures sited on the surface. Pittsburgh's National Bureau of Mines puts on a colorful slide show for selected audiences showing the many hazards of coal mine subsidence. It features shopping center foundations that have collapsed, pipelines

BROADRANGING APPLICATIONS OF THE NAVSTAR GPS

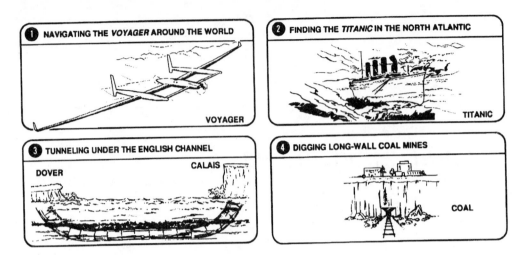

Figure 15.1 The Navstar Global Positioning System has often been employed in creative and interesting ways, a sprinkling of which have appeared on the nightly news. As these four sketches indicate, GPS position-fixing has been adopted for use in connection with such newsworthy projects as the Voyager's around-the-world flight, the construction of the English Channel tunnel, and the location of the Titanic deep under North Atlantic waters.

snapped in two, bridges broken, and electrical transmission towers tilted in tangled disarray.

Subsidence rates for long-wall coal mines can be measured by teams of surveyors who reestablish the locations of surface-mounted benchmarks every week or so. Interferometry receivers can measure subsidence rates continuously, radioing the results to nearby calibration stations so that the data can be processed in real time by powerful computers.

Dinosaur Hunting with the GPS

In the early 1980s Dr. Roy P. Mackal, a research biologist at the University of Chicago, and his colleague, Dr. Herman Regusters, an engineering consultant at Cal Tech's Jet Propulsion Laboratory, encountered tantalizing accounts of a strange dinosaur-like creature, the "mokele mbembe," said to inhabit the Belgian Congo.

According to descriptions circulated by the local pygmies, the giant animal, which spent most of its time immersed in swampy waters, had a "reddish-brown body, a long neck, a relatively small head, and a long, powerful tail." When Mackal asked the natives to examine realistic sketches of many different kinds of animals, they agreed that the mokele mbembes most closely resembled the brontosaurus—the largest land animal ever to stalk the earth (see Figure 15.2). The mokele mbembes' favorite hangout was

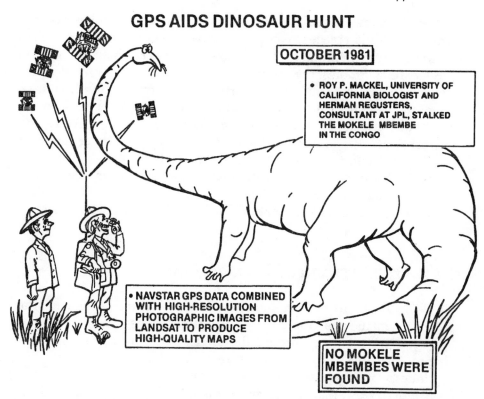

Figure 15.2 In 1981 Roy P. Mackel and Herman Regusters decided to search the Belgian Congo for "mokele mbembes," dinosaur-like creatures that had been reported by missionaries, explorers, tourists, and the local pygmies. Such a large creature, they concluded, would have to live in swampy areas because of the great difficulty it would encounter in attempting to support its own body weight. So they spent hours studying Landsat images to pinpoint the swampiest regions in that portion of the Belgian Congo. Mackel and Regusters attempted to borrow one of the DoD's GPS receivers to guide them on their way. No units were available. They went anyway, but did not find any mokele mbembes.

a big lake frequented by the local fishermen who had, according to one contemporary account, erected a barrier across one of the rivers feeding the lake in order to keep the big, hungry beasts away.

Dr. Mackal became even more intrigued with the possibility of making a careful search when he learned that several western observers had also picked up traces of the oversized animals lurking about in the dense jungles of the Belgian Congo. In the 1770s, for instance, French missionaries had reported seeing "the tracks of a clawed animal the size of an elephant," and two centuries later two separate German expeditions "hinted at the existence of the elusive monster."

When he ventured to the Congo in 1982, Roy P. Mackal carried video cameras and a packet of sonar devices designed to monitor any subtle underwater movements of the local mokele mbembes. He was hoping to find them paddling around in the swollen streams during the rainy season when water levels would likely rise enough for their families to venture out from their hiding places in the jungle and come downriver in search of food.

Mackal was planning to take a Navstar receiver to help guide his expedition to the swampiest regions he could find on false-color images picked up by earth resources satellites. Unfortunately, borrowing a Navstar receiver from the DoD turned out to be an impossible dream, because all the ones they had purchased were tied up in military test programs, mostly at Yuma, Arizona. Mackal journeyed to the Congo anyway, determined, but ill-equipped. He found no mokele mbembes.

Thus, the conclusion seems obvious: if only a Navstar receiver had been available, the zoos of the world would, today, probably be filled with gigantic brotosaurus-like creatures supposedly long since extinct.

Guiding Archaeological Expeditions

In 1984, Oklahoma businessman Ron Frates ran across a colorful article in a flight magazine describing the painful tribulations of a gang of archaeologists hacking their way through the jungles of the Yucatan Peninsula searching for the ruins of an ancient Mayan civilization.

Frates was impressed with their tenacity. But their basic approach, he concluded, was an anachronism in the age of space. So he began to pour over a collection of Landsat images, hoping to piece together subtle clues that might reveal the remains of ancient Mayan settlements hidden under the dense jungle foliage just north of Guatemala and Belize. Soon he discovered, to his delight, that ancient Mayan villagers had skinned back the earth to form more than 100 square reservoirs at least 13 centuries before the images were made from space. After checking and rechecking his observations, Frates recruited a few close friends to share his adventure of a lifetime. He then packed up a bundle of Landsat pictures, a signed permit from the Mexican government, and a Navstar receiver he had managed to borrow from Raytheon.

On the first day over the Yucatan, Frates and his colleagues found nothing but big, straight trees. But, early on the second day, they spotted a continuous expanse of low rock walls carefully constructed by ancient Mayan settlers intent on farming the land. One of the stone walls stretched on for nearly 40 miles. These stone enclosures, which were entirely unexpected, presented Frates and his crew with tantalizing clues concerning the daily life-styles of the Mayan villagers 65 generations before.

Until that time most professional archaeologists were convinced that the area's ancient residents had fed their families using slash-and-burn agriculture, a primitive form of cultivation still practiced in many Third World countries. When slash-and-burn agriculture is widely used, land is seldom in short supply, so enclosing the fields with big heavy rock walls would have been a waste of time. The walled farming plots thus pointed toward a much larger Mayan population than had been previously suspected, structured in keeping complex social institutions and centralized control. Some experts are now convinced that, at their peak, as many as 15 million Mayan villagers may have scratched out a living in tree-shrouded areas of the Yucatan.

Guided by his trusty Navstar receiver, Ron Frates and his space-age archaeologists swooped down on many of the promising sites they had picked out in the Landsat images. One of them was apparently Oxpermul, a settlement that had been discovered in the 1930s and then lost again because in those early days archaeologists had no reliable method for marking its location accurately enough so they could come back again another day.

Another high point of the expedition came a day later, when they flew over an astonishing sight: the top of a stone pyramid peeking up through the jungle canopy. Using an old automobile hoist, his colleagues lowered Ron Frates down onto the smooth, flat rocks. At another jungle-covered site there was no place to land, so they quickly reeled two crew members down to clear a crude helicopter landing site. Unfortunately, their chain saw broke, so they had to scamper back up again.

Ron Frates and his colleagues were enormously pleased with the results of his hastily organized archaeological expedition. "We were able to map the extent of the Mayan civilization on the Yucatan in about five days," he later concluded. "Working on foot it would have taken at least one hundred years."

Tracking Hazardous Icebergs

In 1912 the "unsinkable" British luxury liner Titanic slurped its way down toward Davy Jones Locker when it suddenly collided with an Arctic iceberg. That incident received worldwide publicity, but, actually, it represented only one of many tragic encounters between hazardous sea ice and maritime vessels.

Each year 10,000 icebergs form in the Northern Hemisphere. About 5,000 of them reach the open ocean, and, on average, 300 pass below 48 degrees north latitude where they become a major hazard to North Atlantic shipping. During the iceberg season, which lasts about five months, shipping routes are lengthened as much as 30 percent so that large vessels can avoid the worst concentrations of floating ice. Gangling oil platforms in northern waters are also at hazard and, on occasion, they must be abandoned when icebergs come into their vicinity.

Of course, no region in the North Atlantic is entirely free from iceberg hazards. Consequently, Arctic icebergs are constantly tracked by the International Ice Patrol, which is managed by the U.S. Coast Guard, but financed by various maritime nations operating in the North Atlantic. The International Ice Patrol sends ships and airplanes into the North Atlantic to locate as many icebergs as possible. Targets of opportunity are also reported by commercial and military vessels. Six radio stations broadcast iceberg warnings twice each day so that ships in the area can safely adjust their movements.

Efforts have been made to track icebergs with electromagnetic sensors onboard orbiting satellites, but, so far, the results have not been entirely satisfactory. Part of the difficulty stems from the fact that, from space, snow-covered Arctic islands closely resemble large icebergs. Moreover, a floating iceberg is difficult to distinguish from one that is landlocked either permanently or intermittently. Shipboard radar devices are also surprisingly ineffective. On the average, an iceberg has only about 1/60th the radar reflectivity of a metal or wooden ship of comparable size. In other words, ice is nature's own natural stealth.[1]

Researchers at the International Ice Patrol have attempted to destroy a few icebergs using various approaches, including thermite explosions, bombing, torpedoing, shelling, ramming, and painting the iceberg with charcoal and lampblack, hoping to increase melting rates. In one test, twenty 1,000-pound bombs were dropped on a 250,000-ton iceberg, but only 20 percent of it was chipped away. Of course, icebergs are often much larger than 250,000 tons—a few are 50 miles long, or even longer.

Because of its high degree of accuracy and its continuous availability, the Navstar Global Positioning System could provide an attractive method for tracking icebergs in North Atlantic shipping lanes. A specially designed receiver would be dropped onto the iceberg to broadcast its position to nearby maritime vessels so that they could take evasive maneuvers. Actually, orbiting satellites have already been used to track at least a dozen icebergs in the North Atlantic.

In one test (see Figure 15.3), an 80-pound transceiver was delivered by parachute to the surface of a flat, Arctic iceberg. The transceiver was then used to relay crude position coordinates to NASA engineers through the Nimbus 6 weather satellite. Over a 3-month interval, it traveled along an ambling and erratic trajectory across 200 miles of open ocean. On the average, it moved less than 5 miles per day, well below walking speed. About 20 percent of the time, the iceberg was essentially stationary because it was either permanently or intermittently grounded.

The Nimbus weather satellite turned out to be a useful tool for tracking icebergs experientially, but the positioning measurements it provided were not very accurate and were not obtained in real time. Consequently, better methods are required to keep tabs on the 300 icebergs that pass below the 48th parallel in an average year. GPS tracking is a definite possibility. The GPS can provide continuous, real-time monitoring of the hazardous icebergs that menace North Atlantic shipping lanes. The resulting measurements could be broadcast directly to nearby vessels whose crews, incidentally, could later pick up the floating receiver when the iceberg melts. By returning it to the International Ice Patrol, they could earn an appropriate reward.

---

[1]One conclusion seems obvious; stealth airplanes should be constructed from thin sheets of ice!

ICEBERG TRACKING

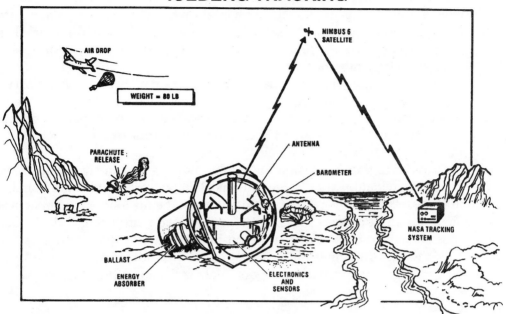

Figure 15.3 So far, American oceanographers have parachuted space-age navigation transmitters onto at least a dozen large Arctic icebergs. Once in place, the transmitter broadcasts tracking pulses through orbiting satellites to help researchers reconstruct the complicated trajectory of the iceberg to within a few thousand feet. With Navstar navigation techniques, position-fixing accuracies will be improved and real-time warnings can be provided to any ships in the vicinity of a hazardous iceberg. Someday this approach may help improve the safety of shipping crews and allow vessels traversing North Atlantic waters to reach their intended destinations faster and with less consumption of precious fuel.

Offshore Oil Exploration

"Oil is the opium of the West" declared British Commentator Kenneth Clarke. To satiate that nagging Western addiction, global demand for petroleum has been trending upward. And, yet, despite extensive exploratory drilling, total known reserves in America's lower 48 states have been declining, apparently because there is not much onshore oil left to discover. For this reason, oil companies have been concentrating their exploratory efforts offshore, where they are hoping to locate major new petroleum deposits. Promising regions on the continental shelf total approximately 6 million square miles; roughly twice the area of the lower 48 states.

Successful offshore oil exploration depends on the accurate reconstruction of three-dimensional subterranean geological structures. Information for the necessary geological maps is gathered by the crews of oil exploration vessels who trigger seismic pulses that travel down through the water and reflect off the subsurface geological layers. When the technicians computer-process the resulting data, they must be able to determine the position of the exploration vessel each time it triggered a seismic pulse. When shore-based navigation beacons are visible, navigation requirements can be met with relative ease, but accurate positioning becomes much more difficult when the ship is situated below the horizon with respect to its shore-based navigation transmitters.

Because positioning accuracies are so important to their mission, most oil exploration vessels are equipped with "integrated" navigation systems costing several hundred thousand dollars each. A typical shipbased system of this type involves a complicated array of equipment jury-rigged into a functioning system.

Firms engaging in offshore oil exploration are anxiously awaiting the accurate navigation and the continuous global coverage available from the full constellation of Navstar GPS satellites. In the meantime, researchers at Shell Canada Products, Ltd., in Calgary, Canada, are using today's GPS constellation to help establish the position of an oil exploration vessel in Arctic waters 180 miles off the coast of Newfoundland (see Figure 15.4). According to their published report, "data collection efficiencies increased

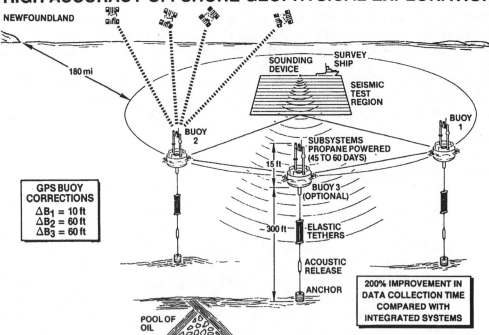

HIGH ACCURACY OFFSHORE GEOPHYSICAL EXPLORATION

Figure 15.4   In this experimental navigation system, three tethered buoys are anchored with elastic tethers to the sea floor 180 miles off the coast of Newfoundland. Each buoy is accurately positioned by picking up signals from the Navstar satellites and, when it must be repositioned, the satellite signals are used again. The buoys are, themselves, navigation transmitters whose pulses help researchers position their off-shore oil exploration ships. According to project engineers, this approach turned out to be twice as efficient as the usual shore-based methods for positioning offshore oil exploration vessels.

by 200-percent compared with missions using more conventional methods of navigation."

Elastic tethers as long as 300 feet were used to anchor transmitting buoys to the continental shelf. This kept them from shifting side-to-side when they were washed by tides and waves. Each buoy transmitted precise positioning pulses to help the seismic exploration ship fix its location with a high degree of accuracy.

The navigation pulses picked up by the ship were recorded on the magnetic tapes for later analysis, using high-speed digital computers. Signals from the GPS satellites helped position the buoys at the beginning of the project, and each time they had to be moved to a new location.

Fixing the Positions of Railroad Trains

The navigation of railroad trains is also being handled on an experimental basis using the Navstar navigation system. The Advanced Railroad Electronics Systems (ARES), for instance, is being developed jointly by engineers and technicians at Burlington Northern Railroad and the Rockwell Collins Air Transportation Divisions. The trains are accurately positioned in real-time, using signals from the GPS satellites. Then their position fixes are relayed through a 900-megahertz radio link to the central dispatchers who carefully monitor and control the movements of the trains, whose images are displayed on a big video screen.

Burlington Northern's technicians are also installing electronic sensors aboard their trains to help keep track of such important performance parameters as oil temperatures, crank-case pressures, fuel quantities, and other measures of locomotive fitness.

Why was GPS navigation selected rather than positioning sensors permanently attached to the tracks? Burlington Northern's director, Edward Butts, recently provided this simple explanation to a magazine reporter: "We drove Rockwell to go with GPS positioning and the reason was economics," he said. "We experimented with track-mounted transponders early on and found a tremendous maintenance burden. The GPS signals are free, and the cost of the positioning determination chips have been coming down." Burlington Northern operates 2,300 locomotives pulling 60,000 freight cars on 23,000 miles of track spanning two Canadian provinces and 25 American states. Their average freight haul spans 760 miles, much of it over single track.

The Advanced Railroad Electronic System is expected to cost about $350 million dollars, with a projected return of $3 for every $1 Burlington invests. With the new navigation and monitoring equipment, control center managers can help speed operations, maintain schedules, and enhance safety by slowing down some cars, speeding up others, and sending still others along entirely different routes. The dispatchers and controllers will also have better information on how to move crews to safety whenever high-speed trains are approaching their work sites.

Automobile Navigation

Automobiles can benefit greatly from the accurate position-fixing signals available from the Navstar GPS. Careful research studies indicate that 6 percent of all distance traveled by non-commercial vehicles is wasted due to inaccurate navigation, together with 12 percent of all automobile travel time. These extra expenditures result from additional vehicle operating costs, additional accidents that would not otherwise occur, and the value of the extra driving time. One study pegged the total cost of inefficient automotive navigation for the United States at $46 billion per year, much of which could be saved through the intelligent use of modern space-based and ground-based radionavigation techniques.

Dead Reckoning Systems

Dead reckoning automotive navigation systems, which fix their positions without relying on any external signals, have been commercially available for nearly ten years. The Etak vehicle navigation system, for instance, uses solid-state flux-gate compasses to determine the vehicle's direction of travel and magnetic wheel-motion sensors to determine its instantaneous speed.[2] These continuous measurements are computer-processed to determine the vehicle's current location to within about 50 feet. Dead reckoning systems of this type build up cumulative position errors that tend to increase relentlessly with time. However, whenever the vehicle makes a right-angle turn or rounds a sharp curve, the Etak computer automatically attempts to match the turn with the map stored in its electronic memory. If it is successful, this gives it a position update to eliminate the cumulative errors.

The engineers at Etak of Sunnyvale, California, introduced their cleverly designed vehicle navigation system in 1985. They have sold thousands of units for about $1,500 each. A video screen mounted in the dashboard displays an electronic map of all the nearby streets, alleys, and highways. A cursor marks the current location of the car and another marks the driver's destination.

As the car travels along the highway, its electronic map rotates and translates automatically, so that the cursor marking the car's location always remains at the center of the screen, and the road ahead always points upward toward the top. Consequently, a driver who looks up at the highway or down at the display sees exactly the same thing. A zoom feature allows the driver

---

[2]Etak is not an acronym. It is the Polynesian word for "navigation."

to shrink or enlarge the display at different levels of magnification. By zooming in or out, various appropriate levels of detail can be brought into view.

The Etak vehicle navigation system has been purchased by several thousand customers, most of whom operate package delivery services. The system works as advertised, but it entails certain serious shortcomings that tend to block its widespread acceptance by the drivers of ordinary private cars:

1. Multiple cassette tapes are required to cover the average city. Four are needed to cover San Francisco, for instance.
2. The displays are in black and white only.
3. Off-the-road navigation is impractical.
4. Manual initialization is required by the driver.
5. The system sometimes gets confused and must be reinitialized. During early tests on the streets of San Francisco, this happened every 70 minutes or so.

Tomorrow's Space-based Vehicle Navigation Techniques

How will future navigation systems be designed to penetrate tomorrow's automotive markets? One promising approach that is, to some extent, already passing into general use, would pick up precise navigation signals from the Navstar satellites to fix its position (see Figure 15.5). A simple dead reckoning system would also be incorporated to provide position updates when buildings and trees block the signals from the Navstar satellites. Full-color maps stored on laser video disks will display all the roads, rivers, and highways in the local area in a striking visual manner. The driver will have access to at least seven different levels of magnification under push-button control.

The onboard computer is used to determine the optimal route to the driver's intended destination, which it marks in color on the video screen. Voice synthesizer chips give the driver clear verbal directions on any necessary lane changes and sharp angle turns as he travels along the road. Current traffic conditions and weather reports will also be data-linked into the system so that it can continually update its routing whenever road conditions or traffic congestion make the optimal route impractical.

Useful information on popular tourist attractions, restaurants, and hotels will be stored on the laser video disks. Access is achieved via touchscreen buttons. If the driver (or a passenger) touches the icon of a particular hotel, full-color pictures of the rooms inside are displayed as the voice synthesizer chip describes the accommodations and quotes current room rates. If the user touches a particular restaurant icon, the menu of the restaurant appears automatically on the screen together with prices, typical dishes, reservation

SPACE-AGE AUTOMOBILE NAVIGATION

Figure 15.5 This sophisticated method of automotive navigation uses a dead reckoning system, together with signals from the Navstar satellites to nail down the position of an ordinary private car. A cursor marking the car's location is superimposed on a full-color map depicting all of the surrounding streets and highways that are stored electronically on a laser video disk. As the automobile travels along its intended route, the electronic map twists and turns so that the cursor marking the car's location always remains at the center of the screen and the road ahead always points upward toward the top. A second cursor marks the driver's intended destination. At the touch of a button, the driver can shrink or enlarge the multi-color maps at any of seven different levels of magnification.

policies, and operating hours as the voice synthesizer chip describes the specialities of the house with appropriate enthusiasm.

Today's Available Automotive Navigation Systems

A few of the cars in Japan have already learned how to navigate using the GPS satellites. In June 1990 Mazda introduced an automotive navigation system with color-coded video maps displayed on a screen in the dashboard. Position-fixing is accomplished by means of the GPS satellites, coupled with computer-based map-matching techniques. Total cost of the system has been quoted at $2,000 per unit, including a full-color liquid crystal display. Laser disks pinpointing all the highways in Japan are already available. They retail for $60 each.

The dash-board mounted receivers for Japan's automotive market are being developed jointly by America's Tremble Navigation and Pioneer Electronics of Japan. Pioneer's marketeers are anticipating a market for 2,000 receivers per month. Pioneer and Tremble started by developing an in-car GPS system for the Mazda Cosmo. Its video screen is also used for monitoring vehicle performance and for the displays associated with the use of the car's cellular telephone system.

Futuristic Applications for Navstar Navigation

Future entrepreneurs are not likely to run out of fruitful ideas on how to make ample amounts of money from Navstar navigation. As Figure 15.6 indicates, the mules that carry tourists down into the Grand Canyon may someday carry a GPS-directed cassette tape recorder strapped to the saddle

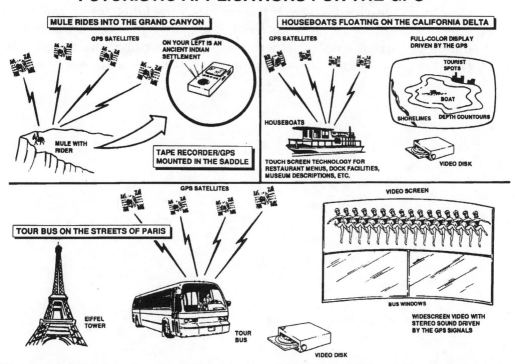

Figure 15.6 Rental markets for GPS receivers could become a billion-dollar-a-year business, especially if creative entrepreneurs manage to couple state-of-the-art audio and video technologies with tomorrow's small and inexpensive Navstar receivers. In these three futuristic applications, signals from the GPS satellites are used in connection with the mules carrying tourists down into the Grand Canyon, the houseboats sailing along the California Delta, and the tour buses zipping along the colorful and exciting streets of Paris.

of each mule. The GPS signals will activate the tape recorder to provide the appropriate commentary whenever mule and rider pass by a tourist attraction of interest. Regardless of the mule's rate of progress, the tape-recorded narration will always match the local surroundings.

Someday soon, the houseboats that carry tourists on the California Delta may also be equipped with GPS-driven video displays with color-coded contours and icons highlighting water depths and navigation hazards. Those same displays will mark the local hotels, restaurants, and other tourist attractions with easy-to-recognize sketches and symbols. The electronic maps will be stored on video disks with touch-screen technology to allow the tourists instantaneous access to restaurant menus, dock facilities, museum descriptions, and the like.

Someday soon, the tour buses whizzing along the streets of Paris will be equipped with widescreen video displays running across the interior of the bus above the driver's head. The images projected on the screen will come from a laser video disk activated by navigation signals from the GPS satellites. As the driver winds through the picturesque streets of Paris along any desired route, the videodisk will be triggered whenever the bus comes near any noteworthy tourist spot. If the bus happens to travel by the Moulin Rouge, for instance, the energetic performers inside will dance a rollicking can-can on the video screen. If it approaches the Eiffel Tower, the video screen will spring to life with movie footage of that brave but overly optimistic Parisian tailor who tried, in vain, to demonstrate that he could fly from the tower by vigorously flapping his artificial homemade wings.

Appendix A _____

Additional Sources of Information

Several professional organizations provide reliable information on the status of the GPS satellites, their current orbital positions, their planned outage intervals, and other related warnings, gossip, and tidbits. Clock status data and precise orbital ephemeris constants are also furnished to the general public by a few government-sponsored and private organizations. Their representatives distribute the desired information in published reports, telephone recordings, and computer bulletin boards.

GPS Information Centers

Information centers devoted to the Navstar GPS are operated by the United States Coast Guard, New Mexico's Holloman Air Force Base, and Global Satellite Software, Inc., in San Jose, California.

The U.S. Coast Guard's Information Center

The Coast Guard representatives broadcast radio announcements over WWV at 13 and 14 minutes after the hour and over WWVH at 43 and 44 minutes after the hour. These regular broadcasts include the Defense Mapping Agency's *Notices to Mariners*, together with status data on the Navstar constellation.

Recorded telephone messages are provided by the GPS Information Center 24 hours a day, with live responses on the same phone lines for 8 hours during the normal working hours of 8 A.M. to 4 P.M. Eastern Standard or Eastern Daylight Saving time. Each announcement includes the number of operational satellites and their PRN (pseudorandom number) designations, together with the most likely dates for planned launches. Prearranged satellite outage announcements are also available from Coast Guard officials.

The telephone number for both live and recorded messages and the two-way fax number are as follows:

(703) 866-3826 (phone)
(703) 866-3825 (fax)

Free computer bulletin boards are available on:

(703) 866-3890 (for modem speeds of 300, 1,200, and 2,400 baud).

Free computer bulletin boards are available on:

(703) 866-3894 (for modem speeds of 4,800 and 9,600 baud).

The communication parameters for all computer bulletin boards are: 8 data bits, 1 stop bit, no parity. Both Bell and CCIH communication protocols are supported by the Coast Guard's service-oriented data link system.

Comments and questions concerning the services available from the GPS Information Center may be directed to:

Commanding Officer
U.S. Coast Guard ON SCEN
7323 Telegraph Road
Alexandria, Virginia 22310-3998
Phone (703) 866-3806 Fax (703) 866-3825
8 A.M. to 4 P.M. Eastern Standard or Eastern Daylight Saving time

Comments and questions about GPS policy matters may be directed to:

COMMANT (G-NRN-2)
U.S. Coast Guard
2100 2nd Street S.W.
Washington, D.C. 20593
Phone (202) 267-0298 Fax (202) 267-4427
8 A.M. to 3 P.M. Eastern Standard or Eastern Daylight Saving time

The Computer Bulletin Board at Holloman Air Force Base

Military personnel at Holloman Air Force Base in New Mexico have taken over the GPS data distribution services formerly originating from the military test range at Yuma, Arizona. Holloman officials provide daily almanacs, observed ranging errors, and current information on the number and locations of the active satellites in the GPS constellation.

A live operator can be reached for assistance on:

(505) 679-1784

Modem services are available from:

(505) 679-1525

The modem operates at full duplex, 8-bit data words, no parity and one stop bit *or* full duplex, 7-bit words, odd or even parity and one stop bit.

For further information contact:

Senior Master Sergeant
Walt Alsleben or Andy Chasko
(505) 679-2151

Global Satellite Software's Computer Bulletin Board

Daily almanacs may be obtained free of charge by computer modem. Write or call:

Glen Siebert
Global Satellite Software, Inc.
5339 Prospect Road, Suite 239
San Jose, California 95129
(408) 252-7490

Modem connections are available on this number:

(408) 252-7358 (with data transmission rates of 1,200 or 2,400 baud).

The Glonass Computer Bulletin Board

Computer bulletin board information relating to the Soviet Glonass constellation of radionavigation satellites is available from:

Jim Danaher
3S Navigation
23141 Plaza Pointe Drive
Laguna Hills, California 92653
(714) 830-3777

the modem phone access number is:

(704) 830-3794

Computer-to-computer transmission rates are available at 1,200 and 2,400 baud.

Precise GPS Orbit Information

Commercially available GPS orbital ephemeris data may be obtained from Western Atlas International. For modem format descriptions, fee schedules, and so forth, contact:

Jim Cain
Manager, GPS Services
Western Geophysical Division
Western Atlas International
3600 Briarpark Drive
Houston, Texas 77042-4299
(713) 964-6345

Military GPS Information Directory

For information on the military benefits and application, policies, and so forth, for the Navstar Global Positioning System, contact:

> USAF Space Division
> 6592 ABG/DAD
> Documentation/Publications Branch
> P.O. Box 92960
> Worldway Postal Center
> Los Angeles, California 90009

For satellite status information call:

> (805) 866-5948

or write to the following address:

> HQ USAF/RDSD
> Washington, D.C. 20330

For information on NATO applications and NATO nation policies relating to the Navstar GPS, contact any of the following NATO Centers:

Belgium
MOD Belgium
Air Force Staff/Avionics Branch (VDT/B)
Quartier Reine Elisabeth
1 Rue D/Evere
1140 Brussels, Belgium

Canada
National Defense Headquarters
Attn: DCDS/DAR 3
101 Colonel By Drive
Ottawa, Ontario K1A OK2
Canada

Denmark
CHOD Denmark
Attn: MN 109
PO 202
DK-2950
Vedback, Denmark

France
STCAN
8 Boulevard Victor
75732 Paris, France

Federal Republic of Germany
GMOD Rue V1 3

Postfach 1328
5300 Bonn
Federal Republic of Germany

Italy
Stato Maggiore Aeronautica
Vinle dell Universita 4
00185 Rome, Italy

The Netherlands
Office of Material Development
Kalvermarkt 28
Postbus 20701
2500ES The Hague, The Netherlands

Norway
Hq Defense Command
CANDE
Oslo MIL
Oslo l, Norway

United Kingdom
MODUK PE (A D/A Radio 2)
St. Giles Court
1–13 St. Giles High Street
London C211 8L1
United Kingdom

GPS *Information with a European Flavor*

The United Kingdom, The Netherlands, and Norway all have their own dedicated electronic information services dealing with the Navstar navigation system.

The United Kingdom

For information on the Navstar GPS in the United Kingdom contact:

W. Blanchard
Royal Institute of Navigation
1 Kensington Gore
London SW7 2AT
United Kingdom
44-71-943-6740

A computer bulletin board is operated by The Royal Institute of Navigation, but full access is available only to members of The United Kingdom Civil Satellite Group. For further information call:

44-602-422111
2,400 baud rate
Protocol: N-8-1

The Netherlands

The Survey Department of Rijkswaterstaad, The Dutch Ministry of Transport and Public Works, operates a computer bulletin board devoted to issues, policies, and information of interest to Navstar users. For further information contact:

Harry Landa or Hans Van der Wal
Rijkswaterstaad
P.O. Box 5023
2600 GA Delft
The Netherlands
31-15-691111

The computer bulletin board can be reached on:

31-15-561959

It operates at rates of 1,200 and 2,400 baud.

Protocols: N-8-1

Norway

The Norwegian Mapping Authority (Statens Kartverk) operates a GPS-oriented computer bulletin board for the citizens of Norway. For further information contact:

> George Preiss or Anne Randi Enger
> Statens Kartverk
> N-3500 Hønefoss
> Norway
> 47-67-24100

Their computer bulletin board can be reached on:

> 47-67-24045

It operates at 1,200 and 2,400 baud.

> Protocols: N-8-1

GPS Clock Behavior

The U.S. Naval Observatory in Washington, D.C. provides status information on the GPS satellites, with current timing data on their onboard atomic clocks. The U.S. Naval Observatory Series 4 Weekly Bulletins are sent through the mail to interested parties.
For further assistance contact:

> Francine Vannicola
> U.S. Naval Observatory
> Washington, D.C. 20392-5100
> (202) 653-1525

The Naval Observatory also operates its own computer bulletin board, with the following access numbers and baud rates:

> (202) 653-0068 1,200 baud
> (202) 653-0155 2,400 baud
> (202) 653-1079 9,600 baud

> Communication parameters: 8 data bits, 1 stop, no parity
> terminate lines with CR/LF

The password is: CESIUM 133

Information for Surveyors

The National Geodetic Information Center provides precise orbital information on the GPS satellites based on tracking data collected by stations in the Cooperative

International GPS Tracking Network. Each data set provides accurate ephemeris information spanning one week measured at 15-minute intervals. The ephemeris data is furnished to interested users on floppy diskettes.

The diskettes can be ordered from:

The National Geodetic Information Center
N/CG174, Room 24
National Geodetic Survey
NOS-NOAA
Rockville, Maryland 20852
(301) 443-8631

The National Geodetic Survey also distributes technical reports comparing the performance capabilities for various types of geodetic surveying systems that use the GPS. Two written reports comparing both hardware units and software modules are currently in print:

- 10-model collection for $3.25
- 3-model collection for $2.00

Checks for the amount of purchase should be made out to *The National Geodetic Survey* and forwarded to this mailing address:

National Geodetic Information Center
Chairman, Federal Geodetic Control Committee
N/GC, WGC-1, Room 1006
NOAA Charting and Geodetic Services
Rockville, Maryland 20852
(301) 443-8631

GPS World Magazine

The GPS has its own magazine! It features colorful, eye-catching graphics and beautifully written articles of interest to the GPS user community. Test results, system developments, practical applications, and policy debates are all featured in this lively and interesting publication. For a free subscription, contact:

GPS World
859 Willamette Street
P.O. Box 10955
Eugene, Oregon 97440-2460
(503) 343-1200

In your cover letter on company stationary, explain why you need the magazine and describe your professional responsibilities.

The Federal Radionavigation Plan

The Federal Radionavigation Plan is jointly published every other year by the Department of Transportation and the Department of Defense. Its primary purpose is to define and resolve policy issues dealing with the various types of civil and military radionavigation systems currently being operated in the United States. Installation plans, phase-out commitments, and user requirements are all discussed in detail, with heavy emphasis on the Navstar Global Positioning System.

The Federal Radionavigation Plan summarizes the salient characteristics of various popular navigation systems now being used by the three branches of the U.S. government and by private citizens groups. The number of users for each system is also tabulated by user category. If you have decided to read and study only one document related to radionavigation and the GPS, this would be a superb selection.

The Federal Radionavigation Plan is distributed by:

The National Technical Information Service
Springfield, Virginia 22161

It is designated by the following alphanumeric characters:

DOT-VNTSC-RSPA-90-3/DOD-4650.4.

Single copies sell for about $24.00.

Appendix B _____

Today's Global Family of User-set Makers

At least 50 different companies around the world make and sell Navstar receivers. The units they produce range from small, hand-held devices weighing only a few ounces to differential navigation base stations weighing 200 pounds or even more. Names, addresses, and personal contacts for the various receiver manufacturers are provided in the following alphabetical listings, which are partitioned into *domestic* and *foreign* marketeers.

Domestic User-set Makers

Allen Osborne Associates
756 Lakefield Road
Building J
Westlake Village, CA 91361
Contact: Skip Osborne
(805) 495-8420
Fax: (805) 373-6067

Ashtech, Inc.
390 Potrero Avenue
Sunnyvale, CA 94086
Contact: Dr. Javad Ashjaee
(408) 737-2400
Fax: (408) 737-2407

Austron, Inc.
P.O. Box 14766
Austin, TX 78761-4766
Contact: Don Mitchell
(512) 251-2313
Fax: (512) 251-9685

Bendix/King
General Aviation Avionics Div.
400 North Rogers Road
Olathe, KS 66062-1212
Contact: John Carocari
(913) 782-0400
Fax: (913) 764-5847

DATUM, Inc.
1363 S. State College Boulevard
Anaheim, CA 92806
Contact: Gary L. Geil,
or Marty Ficken
(714) 533-6333
Fax: (714) 533-6345

Del Norte Technology, Inc.
1100 Pamela Drive
Euless, TX 76040
Contact: Devra Fuller
(817) 267-3541
Fax: (817) 354-5762

Furuno

P.O. Box 2343
S. San Francisco, CA 94083
Contact: Jasper Sipes
(415) 873-9393
Fax: (415) 872-3403

GARMIN Corp.

11206 Thompson Avenue
Lenexa, KS 66219
Contact: Tim Casey
(913) 599-1515
Fax: (913) 599-2103

Global Wulfsberg Systems

2144 Michelson Drive
Irvine, CA 92715
Contact: Randy Lincoln
(714) 851-0119
Fax: (714) 752-0604

Honeywell

P.O. Box 21111, MS: K19B4
Phoenix, AZ 85036
Contact: Cindy Morris
(602) 436-1677
Fax: (602) 436-2252

Interstate Electronics Corp.

1001 East Ball Road
Anaheim, CA 92803
Contact: Robert Snow
(714) 758-4164
Fax: (714) 758-4148

ITT Avionics

500 Washington Avenue
Nutley, NJ 07110
Contact: Larry Peterson
(201) 284-3094
Fax: (201) 284-3334

Koden International

77 Accord Park Drive
Norwell, MA 02061
Contact: Kerry Hohl
(617) 871-6223
Fax: (617) 871-6226

Leica, Inc.

Survey Division
40 Technology Park
Suite 100
Norcross, GA 30092
Contact: Lloyd Penland
(404) 447-6361
Fax: (404) 447-0710

OR

Leica Herbrugg
CH; 9435 Herbrugg
Switzerland
Contact: Peter Jackson
41 (71) 70 33 84
Fax: 41 (71) 70 39 99

Litton Aero Products

6101 Condor Drive
Moorpark, CA 93012
Contact: Abdul Tahir
(805) 378-2039
Fax: (805) 378-2199

Magellan Systems

960 Overland Court
San Dimas, CA 91773
Contact: Sharon Jones
(714) 394-5000
Fax: (714) 394-7050

Magnavox, APS Co.

2829 Maricopa Street
Torrance, CA 90503
Contact: Aneil Aschee
(213) 618-1200
Fax: (213) 618-7001

Marinetek

2239 Paragon Drive
San Jose, CA 95131
Contact: Clyde Jacobson
(408) 441-1661
Fax: (408) 441-0809

Motorola, Inc.

Government Electronics Group
8201 E. McDowell Road
P.O. Box 1417
Scottsdale, AZ 85252
Contact: Steve Sheard
(602) 441-7112, or
Chris Moyer (602) 441-7625
Fax: (602) 441-6702

Odetics, Inc.

Precision Time Division
1515 South Manchester Avenue
Anaheim, CA 92802-2907
Contact: Gary Smith
(714) 758-0400
Fax: (714) 776-6363

Prakla-Seismos AG

Buchholzer Str. 100
D-3000 Hannover 51

Germany
Contact: Heinrich Rehmert
49 (5) 11-64 20
Fax: 49 (5) 11-6 47 68 60

Raytheon Marine

46 River Road
Hudson, NH 03051
Contact: Mike Mitchell
(603) 881-5200
Fax: (603) 881-4756

**Rockwell/Collins International
Military equipment:**

Government Avionics Division
400 Collins Road, NE
Cedar Rapids, IA 52498
Contact: Kenneth Bloom
(319) 395-5662
Fax: (319) 395-1642
  Commercial equipment
Commercial GPS B.U.
P.O. Box 568842
Dallas, TX 75356-8842
Contact: Ray Mathis
(214) 996-5731, or
Becky Meuir (214) 996-5863
Fax: (214) 996-7063

Sercel, Inc.

17155 Park Row
P.O. Box 218909
Houston, TX 77218
Contact: Lynn D. Weems
(713) 492-6688
Fax: (713) 492-6910

SI-TEX Marine Electronics

P.O. Box 6700
Clearwater, FL 34620
Contact: Ted Bodtman
(813) 535-4681
Fax: (813) 530-7272

Sokkia

9111 Barton Street
Overland Park, KS 66214
Contact: Norm Whitted
(913) 492-4900 x155
Fax: (913) 492-0188

Sperry Marine, Inc.

1070 Seminole Trail
Charlottesville, VA 22906
Contact: Jack Roeber
(804) 974-2000
Fax: (804) 973-6529

Stanford Telecommunications

2421 Mission College Boulevard
Santa Clara, CA 95050
Contact: Frank Charles
(408) 748-1010
Fax: (408) 980-1066

STC Defence Systems

Navigation Systems Division
London Road
Harlow, Essex CM17 9NA
United Kingdom
Contact: Ian D. Cosh
44 (279) 29531
Fax: 44 (279) 635289
OR
591 Camino de la Reina
Suite 428
San Diego, CA 92108
Contact: Jerry Miller
(619) 295-5182
Fax: (619) 692-0123

Tecom

9324 Topanga Canyon
Chatsworth, CA 91311
Contact: Wayne Englen
(818) 341-4010
Fax: (818) 718-1402

Texas Intstruments

6600 Chase Oaks Boulevard
M/S 8449
Plano, TX 75086
Contact: Frank Houzvicka
(214) 575-4057
Fax: (214) 575-3762

3S Navigation

23141 Plaza Pointe Drive
Laguna Hills, CA 92653
Contact: Heather McCue or
Jim Danaher
(714) 830-3797
Fax: (714) 830-8411

TRAK Systems Division

TRAK Microwave Corp.
4726 Eisenhower Boulevard
Tampa, FL 33634
Contact: Jack McNabb or
Pete Lopez
(813) 884-1411
Fax: (813) 886-2794

Tremetrics

2215 Grand Avenue Parkway

Austin, TX 78728
Contact: Charles Waldridge
(512) 251-1400
Fax: (512) 251-1596

Trimble Navigation, Ltd.
Survey and Mapping Division
585 N. Mary Avenue
Sunnyvale, CA 94086
Contact: Roger Betz
(408) 737-6915
Fax: (408) 737-6074

True Time, Inc.
3243 Santa Rosa Avenue
Santa Rosa, CA 95407
Contact: Jeff McDonald

(707) 528-1230
Fax: (707) 527-6640

Welnavigate Inc.
675 Bonwit Place
Simi Valley, CA 93065
Contact: Sab Ifune
(805) 583-1733
Fax: (805) 583-0046

Western Geophysical Division
Western Atlas International, Inc.
3600 Briarpark Drive
Houston, TX 77042-4299
Contact: Steve Swarts
(713) 974-3194
Fax: (713) 964-6372

Foreign User-set Makers

Canadian Marconi Company
2442 Trenton Avenue
Montreal, Quebec H3P 1Y9
Canada
Contact: Jim Bruce
(514) 341-4182, or
Henry Schlachta (514) 340-3043
Fax: (514) 340-3016

GEC-Plessey Avionics Ltd.
Martin Road, West Leigh,
Havant, Hants PO9 5DH
United Kingdom
Contact: Chris Moyle
44 (0705) 493306
Fax: 44 (705) 493604

Honeywell-ELAC-Nautik GmbH
Westring 425-429
D-2300 Kiel 1
Germany
Contact: Gunnar Wiesner
Fax: 49 (431) 883493

Geotronics AB
Geodimeter Division
Box 64
S-18211 Danderyd
Sweden
Contact: Björn Österlund
46 (8) 753 01 40
Fax: 46 (8) 7532464 / 6305
  OR
55 Leveroni Court

Novato, CA 94949
Contact: Paul Hahn
(415) 883-2367
Fax: (415) 883-2532

Japan Radio Co., Ltd.
Akasaka Twin Tower
17-22, Akasaka 2-Chome
Minato-Ku, Tokyo 107
Japan
Contact: Shin Matsuo
81 (33) 584-8838
Fax: 81 (33) 584-8878

Navstar Ltd.
Royal Oak Way
Daventry, Northants
NN11 5PJ
United KIngdom
Contact: Anthony R. Pratt
44 (327) 79066
Fax: 44 (327) 71116
  OR
Navstar Electronics, Inc.
1500 N. Washington Boulevard
Sarasota, FL 34236
Contact: Patrick Davis
(813) 366-6338
Fax: (813) 366-9335

NEC Corp.
1–10, Nisshincho
Fuchu City, Tokyo 183
Japan

Contact: Shoji Miyazawa
81 (423) 33-1183
Fax: 81 (423) 33-1866

Nov Atel Communications Ltd.

1020 64th Avenue, NE
Calgary, Alberta
Canada T2E 7V8
Contact: Brad Timinski
(403) 295-5053
Fax: (403) 295-0230

Racal Research Limited

N. Wetlands Industrial Estate
Molesey Road
Walton-on-Thames
Surrey KT12 3PL
United Kingdom
Contact: Peter Diederich
44 (932) 228851
Fax: 44 (932) 229805

Rauff & Sorensen Shipmate

Ostre Alle 6
9530 Stovring
Denmark
Contact: Odin Sletten
45 (98) 37 34 99
Fax: 45 (98) 37 38 07
OR
Robertson-Shipmate
400 Oser Avenue
Hauppauge, NY 11788
Contact: Tom Manzari
(516) 273-3737
Fax: (516) 273-3270

Rokar International, Ltd.

Science Based Industry
 Campus

Mount Hotzvim
P.O. Box 3294
Jerusalem 91032
Israel
Contact: Carmel Sofer
972 (2) 822222
Fax: 972 (2) 866238

SEL (Standard Electrik Lorenz AG)

Defence and Aerospace
 Division
Aerospace Systems
Lorenzstrasse 10
D-7000 Stuttgart 40
Germany
Contact: P. Scholl
49 (711) 821-0
Fax: 49 (711) 869-4004

SONY Corporation

General Audio Div. No. 2
Satellite Comm. Systems
Shibaura Technology Center
1-7-4, Konan, Minato-ku
Tokyo 108
Japan
Contact: Ryuji Oki
Fax: 81 (3) 458-7593

Sextant Avionique

Navigation Systems
 Division
25 rue Jules Vedrines
26027 Valence Cedex
France
Contact: J.P. Lacroix
33 (75) 798511
Fax: 33 (75) 552250

Detailed tabulations highlighting the performance characteristics of the various types of GPS receivers made and sold by these 50 companies are available from *Navtec Seminars* in Alexandria, Virginia. For each type of receiver, the tabulations provide the model number, number of channels, primary applications, power rating, weight, size, accuracy, and list price. Photographs of selected receivers are sprinkled among the tabulations. Navtech's booklet ranges over 50 pages of densely packed information. It can be obtained from:

Navtec Seminars, Inc.
2775 Quincy Street, Suite 610
Arlington, Virginia 22206-2204
Phone (703) 931-0500
Fax (703) 931-0503

The booklets, which sell for about $20 a copy, are frequently updated to keep the information they contain fresh and relevant.

Appendix C _____

Navigation-related Clubs and Organizations

Joining one or more of the world's many navigation-related clubs and organizations can be a stimulating and enjoyable way to gather additional information on the Navstar GPS, the Glonass, and various other radionavigation systems. Your active participation in such an organization can also help you make important business and professional contacts. The organizations in this listing welcome members with a variety of backgrounds, talents, and interests. If you write or call, most of them will send free brochures describing their major aims, activities, meetings, and publications. Use the resulting materials as a guide as you search for the perfect organization to join. The American Institute of Aeronautics and Astronautics, The Institute of Navigation, and The Royal Aeronautical Society have very different platforms (some of which are only peripherally related to navigation), but they are all interesting organizations to join.

American Astronautical Society (AAS)
c/o University of Colorado
Campus Box 423
Boulder, Colorado 80309-0423
Phone: (703) 866-0020

American Institute of Aeronautics and Astronautics (AIAA)
370 L' Enfant Prominade SW
Washington, D.C. 20024
Phone: (202) 646-7400

British Interplanetary Society
27129 S. Lambeth Road
London, SW8 1SZ
England
Phone: 071-735-3160
Fax: 071-820-1504

Deutsche Gesellshaft fur Ortung
un Navigation e.v.
Pempelforter Strasse 47
D-4000 Dusseldorf
Federal Republic of Germany
Phone: 49-211-369909
Fax: 49-211-351645

International Astronautics Federation
250 Rue Saint Jacques
F-75005 Paris
France

Institute of Astrophysics
University of Leige
5 Avenue D/ Cointe
B-4200 Cointe-Ougree
Leige, Belgium
Phone: 32-41-529980
Fax: 32-41-527474

Institute of Electrical and Electronics Engineers (IEEE)
445 Holes Lane
Piscataway, New Jersey 08855-1331
Phone: (908) 981-0060

Institute of Navigation (ION)
1026 16th Street NW, Suite 104
Washington, D.C. 20036
Phone: (202) 783-4121

Institute of Space
Instituto Italiano Di Dritto
Spaziale 251 via Giulia
Rome, Italy

Planetary Society
65 N. Catalina Avenue
Pasadena, CA 91106
Phone: (818) 793-5100

Royal Aeronautical Society
4 Hamilton Place
London W1V 0BQ
England
Phone: 071-499-3515
Fax: 071-499-6230

Royal Institute of Navigation
1 Kensington Gore
London SW7 2AT
England

Appendix D _____

Navigation-related Magazines and Periodicals

Navigation-related magazines and periodicals will help you gain timely and accurate information on the emerging trends and political agendas associated with radionavigation in general, and the Navstar GPS in particular. *GPS World and Navigation: Journal of The Institute of Navigation* carry frequent articles on the GPS satellites, their many applications, and their worldwide user base. The other periodicals included in the following list deal with a number of space- and technology-related subjects from a variety of fresh perspectives.

Aerospace America
370 L' Enfant Prominade SW
Washington, D.C. 20024
Phone: (202) 646-7400

Air and Space
Smithsonian Institution
900 Jefferson Drive
Washington, D.C. 20560
Phone: (202) 287-3733

Aviation Week and Space Technology
McGraw-Hill Inc.
1221 Avenue of the Americas
New York, NY 10020
Phone: (212) 512-2000

Challenge
Ball Aerospace Systems Group
Attn: Marketing Communications
P.O. Box 1062
Boulder, CO 80306-9818
Phone: (303) 939-4000

GPS World
859 Willamette Street
P.O. Box 10955
Eugene, OR 97440-2460
Phone: (503) 343-1200

Navigation: Journal of The Institute of Navigation
815 15th Street, N.W., #832
Washington, D.C. 20005
Phone: (202) 783-4121

The Journal of Navigation
The Royal Institute of Navigation
1 Kensington Gore
London SW7 2AT
Phone: 01-589-5021

Scientific American
415 Madison Avenue
New York, NY 10017
Phone: (212) 754-0550

Glossary

Air Traffic Control   A collection of techniques used in safety controlling the flow of airplanes between airports.

Almanac Constants   A crude set of parameters similar to the more precise ephemeris constants used in approximating the orbits of the satellites in the GPS constellation.

Antenna   A resonant device that picks up faint radio signals and feeds them into a receiver.

Atomic Clock   A highly precise timekeeping device whose inherent stability arises from the quantam mechanical behavior of a particular gaseous element, such as cesium, hydrogen, or rubidium.

Attenuation   Any reduction in the strength or quality of a radio signal due to an intervening medium, such as foliage, the ionosphere, or the atmosphere. The intervening medium distorts, reflects, and refracts the signal as it passes through.

Avionics   Any of the various electronic systems carried onboard an airplane.

Bent-pipe Navigation (*see also* Translator)   A radionavigation technique in which modulated electromagnetic waves are sent from a craft to a distant transmitter for immediate free-space relay to a distant computer processing facility where the navigation solution is performed.

Bit   A binary 1 or a binary 0.

Booster Rocket   A large multistage rocket capable of hurling satellites into space.

Broadcast   Any transmission of modulated electromagnetic waves to distant receivers spanning a broad-ranging geographical area.

Byte   An adjacent sequence of binary digits (usually eight in number) used in representing a single number, letter, or punctuation mark.

C/A-code (*see also* P-code)   A satellite-unique sequence of pseudorandom binary pulses transmitted by a GPS satellite at a chipping rate of 1 million bits per second.

Carrier Tracking Loop (*see also* Code Tracking Loop)   The electronic feedback control loop that allows a GPS receiver to generate and match the electromagnetic carrier waves arriving from a particular GPS satellite. Carrier wave matches from four or more satellites allow the receiver to determine its current velocity.

Carrier Wave   A sinusoidal electromagnetic wave usually, but not always, modulated with information.

CEP (Circular Error Probable)   The radius of a circle that contains 50 percent of all the randomly varying statistical samples occupying a two-dimensional region.

Cesium Clock   An atomic clock whose working element is gaseous cesium.

Chipping Rate   The rate at which an electronic circuit produces binary digits. The GPS satellites produce and transmit C/A- and P-code signals with chipping rates of 1 million and 10 million bits per second, respectively.

Circuit   A complete closed electrical pathway that controls the flow of electrons or other submicroscopic charged particles.

Code Tracking Loop (*see also* **Carrier Tracking Loop**)   The electronic feedback control loop that allows a GPS receiver to generate and match the pseudorandom C/A- and/or P-code pulse trains arriving from a particular GPS satellite. Code matches from four or more satellites allow the receiver to fix its position.

Common-view Mode   A specific time synchronization method in which the GPS receivers at two distant sites have direct line-of-sight access to the same satellite at the same time. Once they pick up the timing pulses from the satellite, they then exchange time-offset information with one another through separate communication channels.

Constellation   Any collection of similar satellites designed to provide multiple coverage or multiple redundancy.

Continuous-tracking Receiver (Multichannel Receiver)   A receiver that tracks four or more GPS satellites simultaneously using four or more dedicated, parallel tracking channels. A multichannel receiver of this type gains continuous access to the pseudo-random binary codes, the L-band carrier waves, and the 50-bit-per-second data stream from all the satellites being tracked.

Data Encryption   The process of modifying a binary pulse train so that unauthorized users cannot extract its full meaning, while authorized users (who are supplied with encryption keys) can gain complete access.

Dead Reckoning   A navigation technique in which a vehicle's current position is estimated by numerically integrating a sequence of measured velocity and/or acceleration increments.

Dependent Surveillance Technique   Any method of tracking an aircraft that uses the devices and instruments it carries for the direct reporting of the aircraft's current position.

Differential Navigation   A special accuracy-enhancing navigation technique in which two radionavigation receivers exchange information with one another on their current navigation solutions to enhance the positioning accuracy of one with respect to the other.

Doppler Shift   A systematic change in frequency of a carrier wave that results when transmitter and receiver are moving at different velocities.

2DRMS (2 Distance Root-Mean-Square)   A specific statistical measure characterizing the scatter contained in a set of randomly varying measurements spread out on a flat plane. 2DRMS is a circular radius that includes 95 percent of the statistical data samples.

Dual-capability GPS/Glonass Receiver (*see also* **Interoperability**)   Any combined radio-navigation system that determines the user's position by picking up real-time signals from both the American GPS satellites and the Soviet Glonass satellites.

Embedded Inertial Navigation System   An integrated navigation system formed by in-serting a GPS receiver directly inside an existing inertial navigation system.

Ephemeris Constants   A small group of parameters used in defining the orbit of a celestial body or a man-made satellite.

Error Budget   A complete tabulation specifying the statistical errors, their sources, and their magnitudes that, when properly combined, constitute an approximation of the total error that will likely result in a real-world situation.

Etak Vehicle Navigation System   A specific dead reckoning system that uses flux-gate compasses and magnetic wheel-motion sensors, together with map-matching tech-niques, to fix the position of a car or truck. The vehicle's current location is displayed on an electronic map. The word "etak" means "navigation" in the Polynesian language.

FAA (Federal Aviation Administration)   A specific agency of the federal government responsible for guiding and directing America's airplanes in flight, with emphasis on operating efficiency and public safety.

Fast-sequencing Receiver (Multiplexing Receiver)   Any GPS receiver that uses a single channel to sequentially track four or more satellites with a sequencing rate so rapid that the receiver gains essentially continuous access to all the 50-bit-per-second data streams of all of the satellites being tracked.

Federal Radionavigation Plan   A widely circulated radionavigation document published every other year by the Department of Transportation and the Department of Defense. It summarizes the U.S. government's current plans and policies for fostering, maintaining, and phasing out various domestic and international radionavigation systems.

Figure of Merit   A single-decimal digit ranging between 0 and 9 displayed by a military GPS receiver. The figure of merit provides a rough measure of the current navigational error of that receiver.

Foliage Attenuation   Any reduction in signal strength or signal quality resulting from the limbs and leaves of trees situated along the signal's line-of-sight path.

Force Multiplier Effect   Any enhancement to the military strength of a world power resulting from superior technologies as opposed to a direct enlargement in the number of troops fielded, the number of munitions they use, or the explosive power of those munitions.

Frequency (*see also* **Wavelength**)   The average rate at which an electromagnetic wave oscillates.

General Theory of Relativity (*see also* **Special Theory of Relativity**)   A mathematical theory developed by Albert Einstein, defining the systematic changes that occur in length, mass, and time when a moving object or a light beam pass through a strong gravitational field.

GDOP (Geometrical Dilution of Precision)   Any loss in positioning accuracy resulting from the fact that the various transmitters in a radionavigation system are not optimally situated with respect to the user.

Gigahertz   One billion cycles per second.

Glonass   A specific spaceborne radionavigation system financed by the Soviet Commonwealth consisting of 21 satellites plus 3 active on-orbit spares arranged in three orbital rings, 11,232 nautical miles above the earth.

GPS (Global Positioning System)   A specific spaceborne radionavigation system financed by the Department of Defense consisting of 21 satellites plus 3 active on-orbit spares arranged in 6 orbital rings, 10,898 nautical miles above the earth.

GPS Receiver (GPS User Set)   An electronic device that picks up the modulated signals from four or more GPS satellites, and then demodulates and computer-processes them to obtain a sequence of real-time position, velocity, and timing estimates.

Gravity Wave   Any electromagnetic wave created by the rotation of large, nonsymmetrical celestial bodies. Einstein's theory predicts the existence of gravity waves, but they have not yet been detected unequivocally.

Ground Antennas   The S-band antennas and the associated electronic devices used in uploading the GPS satellites with fresh ephemeris constants and clock-correction factors.

Hertz   One cycle per second.

Hybrid Radionavigation (*see also* **Interoperability**)   Any partnership in which the signals from two or more radionavigation systems, such as GPS and Omega, are used in solving for a user's position.

Hydrogen Maser   An atomic clock whose inherent stability is based on the quantum-mechanical behavior of gaseous atomic hydrogen.

Hyperbolic Ranging (*see also* **Spherical Ranging**)   Any radionavigation technique in which a navigation receiver measures the time-difference-of-arrival of a direct radio signal and a duplicate version of that signal relayed from a second transmitter. When a single time-difference-of-arrival measurement of this type has been completed, the user is known to lie somewhere along a specific hyperbolic line of position.

ILS (Instrumental Landing System)   A specific short-range, ground-based radionavigation system used in landing airplanes at instrumented airports. The Instrumented Landing System, which is gradually being replaced by the Microwave Landing System, keeps the aircraft aligned with the airport runway as it descends along the desired glideslope.

Integrated Navigation System (*see also* **Dead Reckoning**)   An electronic partnership in which two separate methods of navigation work together to enhance their reliability, survivability, and positioning accuracy. Such a device often consists of an inertial navigation system combined with a radionavigation system, such as the Navstar GPS.

Integrating Accelerometer   A device that continuously measures the acceleration of a craft, while numerically integrating its acceleration profile to obtain current velocity, and then integrating again to obtain current position.

Interferometry Receiver (Codeless Receiver)   A specific type of GPS receiver that uses measurements from the L_1 and/or L_2 carrier waves to improve the accuracy of its position estimate relative to a base station.

Interoperability (*see also* **Integrated Navigation System**)   The ability of a navigation receiver to accept and process inputs from two or more different types of navigation systems.

Ionosphere   A set of four roughly concentric layers of charged particles in the earth's upper atmosphere. The ionosphere bends, distorts, and reflects electromagnetic waves.

Iterative Solution   Any numerical approximation that uses a repetitive loop to gradually converge toward a desired solution.

JTIDS Relnav (Joint Tactical Information Distribution System Relative Navigation)   A specific military ground-based communication/radionavigation system that uses time-division multiple-access techniques for message exchange and navigation. JTIDS Relnav is a transportable system being developed by the U.S. Air Force for use in local battlefield areas.

Kalman Filter   A specific type of data combiner used by radionavigation systems in determining an instantaneous position estimate from multiple time-sequenced statistical measurements.

L_1 and L_2   The two specific L-band frequencies transmitted by each Navstar satellite. Two frequencies are employed because their time-of-arrival variations can be used by a military receiver to virtually eliminate navigation errors arising from ionospheric time delays.

Landsat D   A specific earth resources satellite rigged to use a specially designed GPS receiver to fix its position in space.

Line of Position   A partial navigation solution forming a specific well-defined locus of points, such as a circle, hyperbola, or ellipse. The user's actual location is at the intersection of two or more lines of position.

Loran C/D   A specific ground-based radionavigation system operated primarily in coastal areas of the Northern hemisphere. The Loran transmitters broadcast binary pulse trains to allow any nearby receivers to establish their longitude-latitude coordinates.

Marine Chronometer   A precise shipboard timekeeping device invented and perfected by Englishman John Harrison in the middle of the eighteenth century. The marine chronometer can be used in conjunction with the sextant and special ephemeris tables to fix the longitude and latitude of a ship.

Master Control Station   A manned computer processing facility that obtains large collections of pseudo-range measurements from the unmanned monitor stations then uses those measurements in an over determined solution to determine orbital elements and the clock correction factors for the GPS satellites.

Megahertz   One million cycles per second.

Microsecond   One millionth of a second.

Millisecond   One thousandth of a second.

MLS (Microwave Landing System)   A specific short-range, ground-based radionavigation system used in landing airplanes at instrumented airports. The Microwave Landing System, which is gradually replacing the Instrumented Landing System, provides the aircraft with real-time position information that allows it to approach the runway along steep or shallow, curved or segmented trajectories.

Molniya Orbit   A specific elliptical satellite orbit pioneered by the Soviets with a 12-hour period and a 63.4-degree orbital inclination.

Monarch   A specific spaceborne GPS receiver being marketed by Motorola. The Monarch uses interferometry techniques to fix its position in space to a high degree of accuracy.

Monitor Stations   Unmanned facilities at widely separated locations that measure the pseudo-range to each GPS satellite as it sweeps across the sky.

Multi-Year Spacecraft Procurement (Multi-Year Block Buy)   Any satellite procurement contract in which the U.S. government guarantees the manufacturer that it will purchase a large, specific number of satellites over a specified number of years.

Multipath Error   The statistical error in the pseudo-range that results when a portion of the GPS signal is reflected from nearby surfaces, thus smearing and distorting their pseudorandom C/A-and P-code pulses.

Nanosecond   One billionth of a second.

Navigation   The process of fixing the position of a craft and directing that craft from one known location to another.

Navsat   A specific space-based radionavigation system being developed by European space scientists. The Navsat constellation involves six geostationary satellites and nine other satellites in 24-hour elliptical Molyniya orbits inclined 63.4 degrees with respect to the equator.

Omega   A specific ground-based radionavigation system with globally distributed transmitters. The Omega transmissions, which originate in phase, reflect off the ionosphere to provide essentially global coverage with only eight transmitting stations. In its normal operating mode, the Omega receiver positions itself on two intersecting hyperbolas by noting the phase-difference-of-arrival between pairs of carrier waves originating from two different Omega transmitters.

Orbit   The gravity-induced path followed by a satellite or a celestial body as it coasts through space.

P-Code (*see also* **C/A-Code**)   A satellite-unique sequence of pseudorandom binary pulses transmitted by a GPS satellite at a chipping rate of 10 million bits per second.

Perturbation   Any force that tends to distort the simple gravity-induced orbit of a satellite.

Picosecond   One trillionth of a second.

PLRS (Position Location and Reporting System)   A specific ground-based communication/radionavigation system that uses time-division multiple-access techniques for navigation and message exchange. PLRS is a transportable system being developed by the U.S. Army for use in local battlefield areas.

Point Solution   Any instantaneous navigation solution that uses current pseudo-range measurements from four or more GPS satellites to obtain a position fix, without resorting to time averaging or Kalman filtering.

Pseudo-range   The false range between a GPS satellite and a particular user, as determined by multiplying the observed signal travel time by the speed of light. The pseudo-range does not equal the true range because the user set clock is not synchronized with respect to the atomic clock in the satellite.

Pseudo-satellite   A radionavigation transmitter that sits on the ground and transmits its own C/A- and/or P-code pulse sequences similar to the ones being transmitted by the GPS satellites. Pseudo-satellites can extend the coverage area of the normal GPS constellation and enhance its navigation accuracy.

Quartz Crystal Oscillator (*see also* **Voltage-controlled Crystal Oscillator**)   A tiny slab of quartz machined to precise dimensions so that it will oscillate at a particular frequency with high fidelity.

Radionavigation   The use of radio transmissions to fix the position of a craft and direct that craft from one known location to another.

Range Instrumentation   The equipment modules used in measuring performances, positions, events, times, and so on, of the ordinance devices at a military test range.

Repeating Ground Trace   Any locus of subsatellite points that traces out the same longitude-latitude history repeatedly.

Ring Laser Gyro (*see also* **Inertial Navigation System**)   A specific type of inertial navigation system that measures its rotation rates by counting the alternate bands of dark and light (interference fringe patterns) that are created when two counter-rotating laser beams are sent around its closed optical cavity.

Rubidium Clock   An atomic clock whose working element is gaseous rubidium.

Satellite   Any object that orbits the earth or any other celestial body.

Selective Availability (Degradation of Accuracy)   The process of doctoring and distorting the signals coming down from the GPS satellites so that unauthorized users cannot achieve the full military accuracy of the Navstar system.

Sextant   An optical navigation aid with adjustable optics used in measuring the elevation angles of celestial bodies above the local horizon.

Signpost Navigation System   A specific type of ground-based radionavigation system that fixes the position of the user nearest to any of dozens of RF transmitters. The position solution is achieved by relaying the specific identification pulses of the appropriate signpost transmitter to a centrally located computer through a transceiver carried onboard the navigating craft.

Slow-sequencing Receiver (Time-sharing Receiver)   A GPS receiver that tracks four or more satellites, one after the other, using a single tracking channel. A sequential receiver of this type must pause periodically for 30 seconds or more to gain access to at least one 30-second frame of the 50-bit-per-second data stream coming down from each of the GPS satellites being tracked.

Solid-state Device   An electronic valve or amplifier composed of solid, monolithic materials whose electrical properties are controlled by specific impurities purposely inserted into its crystal lattice structure during manufacture.

Spaceborne Receiver   Any onboard GPS receiver used in determining the location of a satellite or a missile in outer space.

Special Committee 104   A specific committee composed of government and industry representatives charged with the responsibility of standardizing the signal formats and the data exchange protocols used by differential navigation transmitters and pseudo-satellites.

Special Theory of Relativity (*see also* **General Theory of Relativity**)   A specific mathematical theory developed by Albert Einstein, defining the systematic changes that occur in length, mass, and time when a moving object approaches the speed of light.

Spherical Ranging (*see also* **Hyperbolic Ranging**)   Any radionavigation technique in which the receiver measures the signal travel time to establish the range to a particular navigation transmitter. When a single ranging measurement of this type has been completed, the user is known to lie on a specific circle of sphere.

Spread Spectrum Signal   Any modulated signal superimposed on an electromagnetic wave in which the number of bits of useful information being transmitted is appreciably less than the bandwidth of the transmission. For the Navstar GPS, the spread spectrum signals provide extra jamming immunity, multipath rejection, navigation accuracy, and robustness.

Star Tracker   An optical device that fixes the position of a missile or a satellite by taking multiple real-time measurements of the apparent positions of known stars.

Starfind   A specific space-based radionavigation system that uses a single geostationary satellite with an antenna consisting of long radiating spokes. Starfind fixes the longitudes and latitudes of ground-based receivers by using narrow-beam frequency transmissions that sweep across the earth repeatedly.

Surveying   The science and art devoted to positioning benchmarks and determining the sizes, shapes, and locations of specific parcels of land.

Time Dilation (Relativistic Time Dilation)   The systematic variation in the rate at which time passes aboard a GPS satellite compared with the rate at which it passes for a GPS receiver on or near the ground. This time variation behaves in accordance with Einstein's special and general theories of relativity due to the differences in gravity and speed between the satellite and the receiver picking up its signals.

Time Synchronization   The process of measuring and adjusting the slight time mismatch between two clocks.

Topex (Topography Experiment)   A specific satellite being developed jointly by the Jet Propulsion Laboratory and the French government. Topex fixes its position by using interferometry techniques to process the signals being picked up from the GPS satellites so that its radar devices can accurately measure the surface contours of the oceans below.

Transit Navigation System (SATNAV)   A specific space-based radionavigation system financed and maintained by the U.S. Navy. Five or six Transit satellites in polar orbits broadcast continuous electromagnetic tones. As the satellite sweeps across the sky from horizon to horizon, the gradual variation in its Doppler shift allows each user on the ground to obtain a single longitude-latitude position fix.

Translator   A simple radionavigation device that picks up the signals from multiple navigation transmitters and then rebroadcasts them to a distant computer, which performs the navigation solution.

Transmission   The broadcast of a modulated electromagnetic wave toward one or more distant receivers.

Triangulation   The process of measuring distances and angles and then using the resulting measurements in trigonometric calculations to determine an area or the length of a line.

User-Equivalent Range Error (UERE)   The overall statistical error in the measured distance between a radionavigation transmitter and a user. The user-equivalent range error arises from position uncertainties in the transmitter, ionospheric and tropospheric delays, multipath reflections, errors in time synchronization, and the like.

Very Long Baseline Interferometry   An astronomical observation technique using several widely separated radio antennas simultaneously observing the same electromagnetic waves to reconstruct a detailed false color image of a specific sector of the sky.

Voltage-controlled Quartz Crystal Oscillator (*see also* **Quartz Crystal Oscillator**)   A quartz crystal oscillator whose frequency can be controlled by making slight adjustments in the voltage driving it.

VOR/DME (VHF Omnidirectional Range/Distance Measurement Equipment)   A specific ground-based radionavigation system used in vectoring airplanes from waypoint to waypoint. The VOR portion uses a narrow scanning beam to give the aircraft the proper azimuth to the next waypoint. The DME portion uses two-way active spherical ranging to give the aircraft the slant range to that waypoint.

Wave Length (*see also* **Frequency**)   The distance in which a sinusoidal oscillation completes one 360-degree sine curve.

WGS-84 (Worldwide Geodetic System 1984)   A specific map coordinate system in which the entire earth is fitted with a particular oblate spheroid, and then partitioned into a standard longitude-latitude coordinate grid.

Worldwide Common Grid   Any global map coordinate system in which particular values for longitude, latitude, and altitude are interpreted in precisely the same way by a broad class of users.

Bibliography

General References

Ackroya, Neil and Robert Lorimer 1990. *Global Navigation: A GPS User's Guide*. London: Lloyd's of London Press.

GPS NAVSTAR global positioning system user's overview. September 1986. Navstar Global Positioning System. A/F Joint Program Office Report YEE-82-0009B. Los Angeles, California.

Helmes, Charles W. and Thomas S. Logsdon. Space-based navigation: past, present, and future. Prepared for presentation at the Annual AIAA Meeting and International Aerospace Exhibit, 1 May 1984, Washington, D.C.

Hurn, Jeff. 1988. *GPS: A Guide to the Next Utility*. Sunnyvale, CA: Trimble Navigation, Ltd.

Logsdon, Thomas S. Industries in space to benefit mankind. Rockwell International report prepared for the National Aeronautics and Space Administration. SP77-AP-0094, 1977.

Logsdon, Thomas S. June 1977. Opportunities in space industrialization. *Journal of Contemporary Business*. pp. 171–184.

Logsdon, Tom. 1988. *Space, Inc.* New York: Crown Publishers.

Spilker, James J. 1977. *Digital Communications by Satellite*. James J. Spilker, Jr., New York: Prentice Hall.

Stansell, Thomas A., Jr. GPS perspectives. Paper read at International Navigation Congress. 2–5 February 1988, Sydney, Australia.

Stechman, Bernadette. 1991. Where in the world am I? *Challenge Magazine*, No. 4.

Toft, Hans. 1987 *GPS Satellite Navigation: New Perspectives in Accurate Navigation*. Stoevring, Denmark. Shipmate.

1990 Federal Radionavigation Plan. 1990. DOT-VNTSC-RSPA-90-3/DOD-4650.4. Published jointly by the Departments of Transportation and Defense. Available to the public through National Technical Information Service, Springfield, Virginia, 22161.

Chapter 1: The Science of Navigation

Coco, David. October 1991. Innovation: GPS-satellites of opportunity for ionospheric monitoring. *GPS World*.

Cooke, Patrick. May 1984. "Look homeward, (electronic) angel." *Science 84*. pp. 75–78.

du Plessis, Roger M. June 1967. Poor man's explanation of Kalman filtering or how I stopped worrying and learned to love matrix inversion. North American Aviation Autonetics Division Company Report.

Gibbons, Glen. April 1991. What in the world!?! *GPS World*.

Logsdon, T. S., and C. W. Helms. The performance capabilities of the navstar space-based navigation system. Paper read at AFCEA 38th International Convention and Exposition, 19–24 June 1984, Washington, D. C.

Logsdon, T. S., and C. W. Helms. Update on the Navstar GPS. June 1983. Technical Marketing Society of American Conference, London, England, and Frankfurt, Germany.

Logsdon, T. S. Satellites bring new precision to navigation. July/August 1982. *High Technology*. pp. 61–66.

Peterson, Benjamin, Keith Gross and Ellen Shirvell. Fall 1990. Analysis of nonlinear Omega receivers. *Navigation: Journal of the Institute of Navigation*.

Wenzel, Robert J. Fall 1988. Omega navigation system—a status report. *Navigation: Journal of the Institute of Navigation*.

Chapter 2: The Navstar Global Positioning System

Georgiadon, Yola and Kenneth D. Doulet. September/October 1990. The issue of selective availability. *GPS World*.

Helms, C. W., and T. S. Logsdon. Space-based navigation: past, present, and future. Paper read at Annual AIAA Meeting and International Aerospace Exhibit, 1 May 1984, Washington, D.C.

Jorgensen, Paul S. Winter 1988–1989. Special relativity and intersatellite tracking. *Navigation: Journal of the Institute of Navigation*.

Klobuchar, John A. April 1991. Innovation: ionospheric effects on GPS. *GPS World*.

Langley, Richard B. May/June 1990. Innovation: why is the GPS signal so complex? *GPS World*.

Logsdon, T. S. and C. W. Helms. Computer simulation and the performance capabilities of the Navstar space-based navigation system. Paper read at IEEE Winter Simulation Conference, 9–11 December 1981, Atlanta, Georgia.

Logsdon, T. A. and C. W. Helms. The Navstar GPS: A statue report. Paper read at 5th Annual Armed Forces Communications and Electronics Association Symposium and Exposition. 24 October 1984, Brussels, Belgium.

Logsdon, T. S., and C. W. Helms. Satellite-based navigation systems. Paper read at EASCON 1981: Electronics and Aerospace Systems Conference, 15–19 November 1981, Washington, D.C.

Sennott, J. W. and D. Pietraszewski. Summer 1989. Experimental measurements and characterization of ionospheric and multipath errors in differential GPS. *Navigation: Journal of the Institute of Navigation*.

Chapter 3: Performance Comparisons for Radio Navigation Systems

Enge, Per K. and James R. McCullough. Winter 1988–1989. Aiding GPS with calibrated Loran C. *Navigation: Journal of the Institute of Navigation*.

Foley, Theresa. Space operations begin using geostar payload. July 25, 1988. *Aviation Week and Space Technology*. p. 55.

Logsdon, T. S. and C. W. Helms. Comparisons between the capabilities of the Navstar GPS and other radionavigation systems. Paper read at EASCON 1981: Electronics and Aerospace Systems Conference, 16 November 1981, Washington, D. C., Rockwell International, SSD 81-0177.

Logsdon, Thomas S. July/August 1984. Satellites bring new precision to navigation. *High Technology*.

Stoddard, Rob. October 1986. Geostar: RDSS on the move. *Satellite Communications*.

Chapter 4: User-set Architecture

Brown, Grover and Patrick Y. C. Hwang. 1991. *Introduction to Random Signals and Applied Kalman Filtering*. New York: John Wiley & Sons.

Hudak, Gregory J. Spring 1986. Navstar global positioning system Collins user equipment: An evolutionary assessment. *Journal of the Institute of Navigation*.

Langley, Richard B. January 1991. Innovation: the GPS receiver—an introduction. *GPS World*.

Lennen, G. R. and Daly, P. Spring 1989. A Navstar GPS C/A-code digital receiver. *Navigation: Journal of the Institute of Navigation*.

MacDoran, Peter F., James H. Whitcomb, and Robert B. Miller. October 1984. Codeless GPS positioning offers sub-meter accuracy. *Sea Technology*.

Small Ceramic microstrip antenna central to surge in new GPS uses. May 6, 1991. *Aviation Week and Space Technology*.

Chapter 5: User-set Performance Comparisons

Clark, R. Kim. July/August 1990. Cost estimating for dual-source procurement of GPS receivers. *GPS World*.

Computer navigation coming to law enforcement. February 1985. *Law and Order*.

DARPA outstanding achievement award. July 1989. *The GPS Quarterly*. Vol. 4, No. 2. Cedar Rapids, Iowa: Rockwell Government Avionics Division.

Global positioning system (GPS) receiver/processor for space applications. August 14, 1987. Rockwell Report T87-790/101.

Pietersen, Otto. April 1991. Two for the road: GPS receiver performance in Scandinavia. *GPS World*.

Chapter 6: Differential Navigation and Pseudo-satellites

Brown, Alison. Fall 1989. Extended differential GPS. *Navigation: Journal of the Institute of Navigation*.

Comsat to Institute Differential GPS Broadcast Service for Gulf of Mexico. July/August 1990. *GPS World*.

Hobbs, Richard R. 1990. *Marine Navigation*. Annapolis, Maryland: Naval Institute Press.

Kalafus, Rudy, et al. Special Committee 104 recommendations for differential GPS service. Proceedings of the 42nd Annual Institute of Navigation Meeting. Seattle, Washington, 23–26 June 1986.

Mack, Giles. October 1, 1991. "Differential GPS and the Skyfix System." Racal Survey Ltd., Surrey, England.

Parkingson, Bradford W. and Kevin T. Fitzgibbon. Optimal location of pseudo-satellites for differential GPS. *Navigation: Journal of the Institute of Navigation*. Winter 1988–1989.

Parkingson, Bradford W., K. P. Schwartz, and Penina Axelrad. Summer 1988. Autonomous GPS integrity monitoring using the pseudorange residual. *Navigation: Journal of the Institute of Navigation*.

Pietraszewski, D., et al. Spring 1988. U.S. Coast Guard differential GPS navigation field test findings. *Navigation: Journal of the Institute of Navigation*.

Stansell, Thomas A., Jr. Spring 1988. RTCM SC-104: recommended pseudosatellite signal specification. *Navigation: Journal of the Institute of Navigation*.

Chapter 7: Interferometry Techniques

Ashkenazi, Vidal and Geraint Ffoulkes-Jones. November/December 1990. "Millimeters over hundreds of kilometers by GPS. *GPS World*.

Axelrod, P. and B. W. Parkingson. Spring 1989. Closed loop navigation and guidance for gravity probe B orbit insertion. *Navigation: Journal of the Institute of Navigation*.

Bertiger, Willy I. and Catherine L. Thornton. Spring 1989. GPS-Based System for Satellite Tracking and Geodesy. *Navigation: Journal of the Institute of Navigation*.

Ewing, Brian D. September/October, 1990. Pseudokinematic GPS for the surveyor. *GPS World.*

Hoar, Gregory J. and Jon E. Maenpa. WM 101: A new tool for geodesy. Paper read at American Geophysical Union Meeting, 9–13 December 1986. San Francisco, California.

Joseph, Keith M., and Paul S. Deem. Precision orientation: A new GPS application. Paper read at ITC Conference, 24–27 October 1983. San Diego, California.

Leick, Alfred. 1990. *GPS Satellite Surveying.* New York: John Wiley and Sons.

Nesb, Inge and Canter, Peter. September/October 1990. GPS attitude determination for navigation. *GPS World.*

Petersen, Carolyn. November/December 1990. Into the Woods with GPS. *GPS World.*

Scherrer, Rene. 1990. The WM GPS primer. Wild Heerbrugg Geodesy Division CH-9435, Heerbrugg, Switzerland.

Stokes, Donald K. and James F. Thompson. January 1991. GPS for railroad corridor surveying. *GPS World.*

Talley, Phillip E. GPS space navigation and pointing system (SNAPS). Paper read at 11th Annual Guidance and Control Conference of the *American Aeronautical Society,* 30 January–3 February 1988, Keystone, Colorado.

Wells, David. 1986. *Guide to GPS Positioning.* Canadian GPS Associates.

Chapter 8: Integrated Navigation Systems

Bielas, M.S. et al. Summer 1988. Test results of prototype fiber optic gyros. *Navigation: Journal of the Institute of Navigation.*

Diesel, John and Eric Guerrazzi. May 1991. Synergistic integration of GPS and INS for civil aviation. *GPS World.*

Hartman, Randolph. Spring 1988. An integrated GPS/IRS design approach. *Navigation: Journal of the Institute of Navigation.*

Hein, Gunter W. et al. Spring 1989. High-precision kinematic GPS differential positioning and integration of GPS with a ring laser strapdown inertial navigation system. *Navigation: Journal of the Institute of Navigation.*

Morrison, Melvin M. Summer 1988. Morrison's qubik inertial measurement unit. *Navigation: Journal of the Institute of Navigation.*

New inertial measurement unit is based on fiber optic gyro. March 11, 1991. *Aviation Week and Space Technology.*

Norling, Brian L. Winter 1987–1988. Superflex: A synergistic combination of vibrating beam and quartz flexure accelerometer technology. *Navigation: Journal of the Institute of Navigation.*

Tazartes, D. A. and J. G. Mark. Spring 1988. Integration of GPS receivers into existing inertial navigation systems. *Navigation: Journal of the Institute of Navigation.*

Donald J. Weber. Spring 1988. A three-axis monolith ring laser gyro. *Navigation: Journal of the Institute of Navigation.*

Warren, Keith. Spring 1981. Electrostatically force-balanced silicon accelerometer. *Navigation: Journal of the Institute of Navigation.*

Chapter 9: Interoperability with Other Navigation Systems

Anodino, T. G. Global positioning system GLONASS. Paper read at 4th Meeting of the Special Committee on Future Air Navigation Systems (FANS), 2–20 May 1988, Montreal, Canada.

Anodino, T. G. Provision of information on navigation satellite failures. ICAO FANS Meeting. 2–20 May 1988, Montreal, Canada.

Dale, S. A. and P. Daly. May 1987. The Soviet Union's GLONASS navigation satellites. IEEE *AES Magazine.*

Dale, S. A., I. D. Kitching, and P. Daly. February 1989. Position-fixing using the USSR's GLONASS C/A-code. IEEE *AES Magazine.*

Daly, Peter. Spring 1991. Progress Toward the Operational Phase of GLONASS. *Navigation: Journal of the Institute of Navigation.*

Higgins, Michael and Hartell Martin. Spring 1988. "Integrated Navigation for Deep Ocean Positioning." *Navigation: Journal of the Institute of Navigation.*

Ivonov, Nicolay E. and Salistchev Vadim. April 1991. GLONASS and GPS: prospects for partnership. *GPS World.*

Klass, Philip J. March 14, 1988. Soviet release of GLONASS data will ease acceptance of NAVSTATS. *Aviation Week and Space Technology.*

Kleusberg, Alfred. November/December 1990. Innovation: Comparing GPS and GLONASS. *GPS World.*

Van Graas, Frank. Summer 1988. Sole means navigation through hybrid Loran C and GPS. *Navigation: Journal of the Institute of Navigation.*

Chapter 10: The Navstar Satellites

The great big garbage dump in the sky. January 1986. *Discover.*

Isakowitz, Steven J. 1991. *International Reference Guide to Space Launch Systems.* American Institute of Aeronautics and Astronautics, Washington, D.C.

Logsdon, Thomas S. December/January 1982. High fliers. *Technology Illustrated.*

Logsdon, Thomas S. June 22, 1985. Orbiting switchboards. *Technology Illustrated.*

Logsdon, Thomas S. 1970. *The Rush Toward The Stars: A Survey of Space Exploration.* Englewood, New Jersey: Franklin Publishing Co.

Logsdon, Thomas S. June 1983. What goes up into orbit doesn't necessarily come down—at least not right away. *Technology Illustrated.*

Logsdon, Thomas S. July/August 1984. Satellites bring new precision to navigation. *High Technology.*

Logsdon, T. S. September 15, 1973. Space debris retrieval. Rockwell International Report SD73-SA-0126.

Chapter 11: Precise Time Synchronization

Dana, Peter H. and Bruce M. Penrod. July/August 1990. The role of GPS in precise time and frequency dissemination. *GPS World.*

Fliegel, Henry F. et al. Fall 1990. An alternate common view method for time transfer with GPS. *Navigation: Journal of the Institute of Navigation.*

Hough, Harold. October 1991. A GPS precise time sampler. *GPS World.*

Klepezynski, W. J. and F. N. Withington. Spring 1988. Operational results of GPS timekeeping. *Navigation: Journal of the Institute of Navigation.*

Logsdon, Tom. The practical benefits of hydrogen masers in space. Paper read at Second International Conference on Frequency Control and Synthesis, 10–13 April 1989, University of Leicester, England.

Vannicola, Vinicio. Fall 1988. Application of time transfer for Omega transmitters using GPS. *Navigation: Journal of the Institute of Navigation.*

Chapter 12: Digital Avionics and Air Traffic Control

Bell, J. C. Summer 1987. Inmarsat and standard C. *Navigation: Journal of the Institute of Navigation.*

Brown, Alison K. Spring 1988. Civil aviation integrity requirements for the global positioning system. *Navigation: Journal of the Institute of Navigation.*

Elson, Banjamin M. July 25, 1983. Transoceanic flight shows GPS uses. *Aviation Week & Space Technology.*

Implementation of a GPS type payload on a geostationary satellite. June 12, 1986. Ford Aerospace and Stanford Telecommunications Report No. RTCA-246-86/SCI/59-25.

Logsdon, T. S. September 15, 1986. FAA applications of the Navstar global positioning system. FY 1986 IR&D Study Report, Project 345. Rockwell Internal Letter GPS-784-86-110.

Logsdon, T. S. and F. Goodale March 8, 1985. Air traffic control and the Navstar GPS. Rockwell International Report SSD-84-0157.

Pilley, H. Robert and Lois V Pilley. October 1991. "GPS aviation and airports—the integrated solution." *GPS World.*

Chapter 13: Geodetic Surveying and Satellite Positioning

Clements, Philip A. September/October 1990. GPS–out of this world. *GPS World.*

Frei, Erwin, Richard Gough, and Fritz K. Brunner. 28 April–2 May 1986. POPS: A new generation of GPS post-processing software. Paper read at Fourth International Geodetic Symposium on Satellite Positioning. Austin, Texas.

Gough, Richard, et al. "The WM 101 and POPS: An investigation in the Swiss Alps." Wild Heerbrugg, 9435 Heerbrugg, Switzerland.

Logsdon, T. S., and C. W. Helms. December 1983. Promising third-world applications of the Navstar global positioning system. Paper read at The African Communication Application of Satellites Conference, Kenya, Africa.

Logsdon, Thomas S. June 1983. What goes up into orbit doesn't necessarily come down—at least not right away." *Technology Illustrated.*

Perez-Urquiola, Manuel Catalan, Manuel Berrocoso Dominguez and Dolores Garcia-Solis Martinez. October 1991. Forging a South American-Antarctica GPS geodetic link. *GPS World.*

Rome, H. James. Fall 1988. Low-orbit navigation concepts. *Navigation: Journal of the Institute of Navigation.*

Smith, Bruce A. April 20, 1979. Four orbiters to get Navstar capability. *Aviation Week & Space Technology.*

Upadhyay, Triveni N., Stephen Cotterill, and Wayne A. Deaton. Autonomous GPS/INS navigation experiment for space transfer vehicle (STV). Paper read at First European Space Agency International Conference on Spacecraft Guidance, Navigation and Control Systems, 4–7 June 1991, Noordwijk, The Netherlands.

Vogel, Shawna. August 1989. "Wobbling World." *Discover.*

Yunck, Thomas P., Slen-Chong Wu, Diun-Tsong Wu, and Catherine L. Thornton. January 1990. Precise tracking of remote sensing satellites with the global positioning system. IEEE Transactions on Geoscience and Remote Sensing.

Chapter 14: Military Applications

Adams, John A. September 1991. Warfare in the information age. *IEEE Spectrum.*

Alexander, George. June 18, 1984. Satellites: Archaeology's newest tool. *Times Science.*

Burgess, Alan. Fall 1989. GPS survivability: A military overview. *Navigation: Journal of the Institute of Navigation.*

Canan, James W. August 1991. A watershed in space. *Air Force Magazine.*

Davis, Harry I. Terrell E. Green, and Seymour J. Deitchman. January 1987. "Tactical air warfare: part II—The Weapons." *Aerospace America.*

Figures that add up to success. April 6, 1991. *Jane's Defense Weekly.*

Georgiadon, Yola and Kenneth D. Doucet. September/October 1990. "Innovation: the issue of selective availability." *GPS World.*

GPS for Tomahawk upgrade. July 1989. *The GPS Quarterly* Vol. 4, No. 2. Cedar Rapids, Iowa: Rockwell Government Avionics Division.

Gulf war's silent warriors bind U.S. units via space. August 1991. *Signal.*

Hoeffener, Carl E. and Joseph T. Stegmaier. June 1991. "Tracking antimissile flight tests with GPS." *GPS World.*

Jacobson, Len. May 1991. "The need for selective availability." *GPS World.*

Keegan, John. April 7, 1991. "The lessons of the Gulf War." *Los Angeles Times Magazine.*

Logsdon, Thomas S. 1984. *The Robot Revolution.* New York: Simon & Schuster.

Logsdon, T. S. and Ashley, J. D. April 26, 1986. Navstar global positioning system applications. Cocoa Beach, Florida: Paper read at International Space Congress.

Military leaders say GPS success in Gulf assures tactical role for satellites. May 13, 1991. *Aviation Week and Space Technology.*

Nordwall, Bruce D. October 14, 1991. Imagination only limit to military, commercial applications of GPS. *Aviation Week and Space Technology.*

Technology: the edge in warfare. September 1991. *IEEE Spectrum.*

U.S. forces praise performance of GPS but suggest improvements. April 22, 1991. *Aviation Week and Space Technology.*

Chapter 15: Civil Applications

Dinosaur hunt. June 1981. *Science Digest.*

French, Robert L. Automobile navigation: where is it going? Paper read at IEEE Position Location and Navigation Symposium, 4–7 November 1986. Las Vegas, Nevada.

French, Robert L. Historical overview of automobile navigation technology. Paper read at IEEE Vehicular Technology Conference, 20–22 May 1986, Dallas, Texas.

Gibbons, Glen. May/June 1990. On track with GPS. *GPS World.*

Honey, S. K., and W. B. Zavoli. A novel approach to automotive navigation and map display. Paper read at Land Navigation and Location for Mobile Applications Conference of the Royal Institute of Navigation, 9–11 September 1985, York, U.K.

Hossfeld, Bernard. October 1991. GPS for vehicle tracking. *GPS World.*

Japanese cars learn to navigate by satellite. May 26, 1990. *The New Scientist.*

Krakwsky, E. J., G. Lachapella, and K. P. Schwartz, 1990. *Assessment of Emerging Technologies for Future Navigation Systems in the Canadian Transportation Sector.* Calgary, Canada: University of Calgary Press.

Logsdon, T. S. and C. W. Helms. Promising civil applications of the Navstar global positioning system. Paper read at National Telesystems Conference, 14 November 1983, San Francisco, California.

Madwell, J. F. and T. S. Logsdon. Space flight opportunities for industry. Paper read at International Space Flight Congress, 12 April 1977, Cape Kennedy, Florida.

Wilbur, Amy. November 1985. From car to satellite. *Science Digest.*

Index